TASTES &

AROMAS

THE CHEMICAL SENSES
IN SCIENCE AND INDUSTRY

EDITED BY
GRAHAM A. BELL
ANNESLEY J. WATSON

Blackwell Science

UNSW PRESS

This volume was published with support from

 The Centre for ChemoSensory Research, the
University of New South Wales, Sydney, Australia

and

 The Cooperative Research Centre for International Food
Manufacture and Packaging Science, Australia

A UNSW Press/Blackwell Science Ltd book

Published by
University of New South Wales Press Ltd
University of New South Wales
Sydney 2052 Australia
http://www.unswpress.com.au

Blackwell Science Ltd
Osney Mead, Oxford OX2 0EL
25 John Street, London WC1N 2BL
23 Ainslie Place, Edinburgh EH3 6AJ
http://www.blackwell-science.com

© Centre for ChemoSensory Research 1999
First published 1999

UK DISTRIBUTORS
Marston Book Services Ltd
PO Box 269
Abingdon
Oxon OX14 4YN
(Orders: Tel: 01235 465500
Fax:01235 465555)

A catalogue record for this title
is available from the British Library
ISBN 0-632-05544-8

US Library of Congress
Cataloging-in-Publication Data is available

National Library of Australia
Cataloguing-in-Publication entry:

Tastes and aromas: the chemical senses
in science and industry.

Includes index.
ISBN 0 86840 769 0
1. Research, Industrial.
2. Science and industry.
3. Taste.
4. Smell.
I. Bell, Graham A.
II. Watson Annesley, J.

573.877

Text design Dana Lundmark
Cover design Di Quick
Printer Everbest, Hong Kong

TASTES & AROMAS

Graham Bell, Director of the Centre for ChemoSensory Research at the University of New South Wales and Adjunct Associate Professor in the School of Physiology and Pharmacology, has been a practising professional scientist in the chemical senses for two decades. He is well known internationally for his research on the fundamentals of smell and taste, the development of novel sensors and the psychology of food.

Annesley Watson is the sensory analyst for Arnott's Biscuits and is the current Chair of the New South Wales division of Australian Institute of Food Science and Technology (AIFST). She has pioneered the application of chemosensory research in the food industry.

CONTENTS

PREFACE

Interest in the chemical senses, in particular smell, taste and pungency, is growing in scientific and industrial circles.

This book offers new information to readers with training and industrial needs. Reviewing the progress being made in chemosensory science and technology shows that this research field does and will provide solutions to industrial problems now and in the future.

Practitioners of sensory research in industry should, by reading this book, be provided with new facts and ideas to contemplate, in the context of their specific needs.

The field of chemical senses covers a wide area of science and technology. We cannot attempt to cover all the known and visible territory, which would take many volumes. Instead we must act as your guide and take you into the minds of prominent individuals in this field and let them describe their small part of it, so that in the end you will know some of the most salient features of the chemosensory landscape and how you might derive benefit from it.

The first four chapters provide introductions to the chemical senses and the important questions currently being studied. We begin with people's earliest thoughts about smell and taste, in the first chapter by Michael Stoddart, a renowned biologist who has made a comprehensive study of chemical communication in humans and other animals. We are reminded of mankind's culturally longstanding fascination with scents, and of their origins in evolution.

Yale University's Linda Bartoshuk and her co-workers then introduce the taste senses, with a perspective on variations that exist in the human population and their implications for food choice, diet and health.

Steven Youngentob gives a clear introduction to the sense of smell, which has for so long been regarded as largely mysterious, but which is now becoming understood at cellular and molecular levels and is generating new technologies, some of which are introduced in this volume.

John Prescott introduces the topic of pungency (the trigeminal sense). Long regarded as too difficult to measure systematically in human beings, the burn of chilli is now a 'hot topic' in food and flavour, with valuable spin-off to the food, confectionery and beverage product developer.

The next five chapters look at value-adding applications of smell and taste research around the world: in Europe (David Lyon), Australia (Jennifer Weller), Asia (Graham Bell and Hae-Jin Song), Japan (Sachiko Saito, *et al.*) and Indonesia (Kerry Easton and Graham Bell).

We then look at two major industrial applications for chemosensory research: the fragrance and wine industries, with chapters by Dragoco Australia's John Lambeth and Ann Noble of the University of California (Davis). These chapters show the central role for human sensory research in developing successful products in billion-dollar global markets.

As human sensory research draws on statistical analysis for sound interpretation of results, we include two chapters illustrating new statistical and design tools for the sensory practitioner by John Best and colleagues and by James Walker and Martin Kendal-Reed.

The following three chapters, by Peter Barry, Susan Sullivan and Brian Key, tell of important fundamental progress in smell and taste research. Anatomical and physiological methods are slowly but surely cracking the process by which chemical compounds are recognised by the mouth and nose and how sensation is transformed into thoughts and feelings. Graham Bell shows in the next chapter that progress on three basic molecular mechanisms of olfaction offers exciting prospects for a range of technologies.

The book concludes with five chapters on artificial chemical sensing. These chapters focus on devices that are now being made to mimic the nose and brain. Alan Mackay-Sim tells of sensor research that will allow miniature robots to follow scent trails and Jelle Atema shows how the crayfish nose can be copied in an underwater 'robo-lobster'. While these may evoke images of demonic toys, they have serious applications in environmental and security industries. Brynn Hibbert reveals how sensors can be used to monitor air and motor vehicle pollution, and Don Barnett describes what is needed for a sensor system to perform a range of valuable tasks in a food factory. David Levy concludes with a clear explanation of artificial neural networks, by which complex information of the kind produced by the senses of smell and taste can be processed and categorised. The brain and the artificial neural net both learn that four hundred volatiles are actually one thing: the smell of chocolate.

Whether you are approaching the science of smell, taste and pungency for the first time or are already involved in teaching, in research or professional practice, we hope all readers will find this book rewarding and will gain useful new perspectives and enhanced appreciation of chemosensory science.

ACKNOWLEDGMENTS

Our sincere thanks go to Roger Edwards and the CRC for International Food Manufacture and Packaging Science for their support of the seminar called 'Sensory Science: Meeting Industry Needs' held in Sydney in November 1996, from which this volume originated; Marilyn Styles of the UNSW Centre for ChemoSensory Research for communicating with authors; John Prescott of the University of Otago, New Zealand, for his insights and assistance in the formulation of the original meeting; and John Elliot of UNSW Press for his valuable commitment to the project. Our thanks also go to Jane Paton and Richard Bell for reading and proofing the final drafts.

One special mention is well deserved: Karyn Weitzner tirelessly desk-edited this volume. We appreciate her skills and her patience.

Naturally, this collection of chapters is nothing without its individual authors. Thanks go to each and every one for giving their time to come to Australia and Sydney and later for their responses when required, particularly as each has updated the original papers for maximum contemporary value to the reader.

CONTRIBUTORS

Editors
Graham Bell and Annesley Watson

CHAPTER 1
D.M. Stoddart
Contact
D. Michael Stoddart
ANARE Chief Scientist
Australian Antarctic Division
Channel Highway
Kingston, TAS 7050, Australia

CHAPTER 2
L.M. Bartoshuk (Yale University School of Medicine), K.E. Cunningham (Yale University School of Medicine), G.M. Dabrila (Yale University School of Medicine), V.B. Duffy (University of Conneticut), L. Etter (Yale University School of Medicine), K.R. Fast (Yale University School of Medicine, University of Conneticut), L.A. Lucchina (Yale University School of Medicine), J.M. Prutkin (Yale University School of Medicine), D.J. Snyder (Florida State University)
Contact
Professor L.M. Bartoshuk
Department of Surgery
Yale University School of Medicine
333 Cedar St.
New Haven, CT 06520-8041, USA

CHAPTER 3
S.L. Youngentob
Contact
Dr S.L. Youngentob
Dept. of Neuroscience and Physiology and The SUNY Clinical Olfactory Research Center
SUNY Health Science Center
Syracuse, NY 13210, USA

CHAPTER 4
J. Prescott
Contact
Dr J. Prescott
Sensory Science Research Centre
University of Otago
PO Box 56
Dunedin, New Zealand

CHAPTER 5
D.H. Lyon
Contact
D.H. Lyon
Head of Department of Sensory Quality and Food Acceptability
Campden and Chorleywood Food Research Association, UK
Gloucestershire, GL556LD, UK

CHAPTER 6
J. Weller
Contact
J. Weller
Sensory Research Manager
Uncle Tobys
Murray Valley Highway
Rutherglen VIC 3685, Australia

CHAPTER 7
G.A. Bell and Hae-Jin Song
Centre for ChemoSensory Research, University of New South Wales
Contact
Associate Professor G.A. Bell,
Centre for ChemoSensory Research
University of New South Wales
Australian Technology Park,
Sydney, NSW 1430, Australia

CHAPTER 8
S. Saito (National Institute of Bioscience and Human Technology (NIBH), Agency of Industrial Science and Technology (AIST), Ministry of International Trade and Industry (MITI), Japan), S. Ayabe-Kanamura (NIBH, AIST, MITI, Japan; Institute of Psychology, University of Tsukuba, Japan); T. Kobayakawa (NIBH, AIST, MITI), Japan); Y. Kuchinomachi (NIBH, AIST, MITI), Japan); and Y. Takashima (Takasago Int. Corp., Japan)
Contact
Sachiko Saito
NIBH, AIST, MITI
Higashi 1-1, Tsukuba,
Ibaraki 305-0046, Japan

CHAPTER 9
K. Easton (Food Science Australia) and G.A. Bell (Centre for ChemoSensory Research, University of New South Wales)
Contact
Kerry Easton
Food Science Australia
P.O. Box 52
Sydney NSW 1670, Australia

CHAPTER 10
J. Lambeth
Contact
J. Lambeth
Fragrance Designer
Dragoco Australia Pty Ltd
PO Box 643
Dee Why, NSW 2099, Australia

CHAPTER 11
A.C. Noble
Contact
Professor A.C. Noble
Department of Viticulture and Enology
University of California
One Shields Ave
Davis, CA 95616, USA

CHAPTER 12
J.C. Walker (FSU Sensory Research Institute, NHMFL), M. Kendal-Reed (Cranio-Facial Centre, School of Dentistry, University of North Carolina) and W.T. Morgan (Research and Development, Bowman Gray Technical Centre, R.J. Reynolds Tobacco Co).
Contact
J. Walker
FSU Sensory Research Institute
NHMFL
Tallahassee, FL 32306, USA

CHAPTER 13
D.J. Best (CSIRO Mathematical and Information Sciences), J.C.W. Rayner (School of Mathematics and Applied Statistics, University of Wollongong) and M. O'Sullivan (CSIRO Mathematical and Information Sciences)
Contact
D.J. Best
CSIRO Mathematical and Information Sciences
PO Box 52
Sydney, NSW 2113, Australia

CHAPTER 14
P.H. Barry (School of Physiology and Pharmacology, University of New South Wales), S. Balasubramanian (School of Physiology and Pharmacology, University of New South Wales) and J.W. Lynch (Department of Physiology and Pharmacology, University of Queensland)
Contact
Professor P.H. Barry
School of Physiology and Pharmacology
University of New South Wales
Sydney, NSW 2052, Australia

CHAPTER 15

S.L. Sullivan
Contact
Dr S. Sullivan
National Institute on Deafness and other Communication Disorders
National Institutes of Health
5 Research Court
Rockville, MD 20850, USA,

CHAPTER 16

B. Key
Contact
Dr B. Key
Neurodevelopment Laboratory
Department of Anatomy and Cell Biology
University of Melbourne
Parkville, VIC 3052, Australia,

CHAPTER 17

G.A. Bell
Contact
Associate Professor G.A. Bell
Centre for ChemoSensory Research
University of New South Wales
Australian Technology Park
Sydney, NSW 1430, Australia

CHAPTER 18

A. Mackay-Sim
Contact
Associate Professor A. Mackay-Sim
School of Biomolecular and Biomedical Science
Griffith University
Brisbane, QLD 4111, Australia

CHAPTER 19

J. Atema
Contact
Professor J. Atema
Professor and Director
Boston University Maritime Program
Marine Biological Laboratory
Woods Hole, MA 02543, USA

CHAPTER 20

D.B. Hibbert
Contact
Professor D. Brynn Hibbert

Department of Analytical Chemistry
University of New South Wales
Sydney, NSW 2052, Australia

CHAPTER 21
D. Barnett
Centre for ChemoSensory Research,
University of New South Wales
& CRC International Food Manufacture
and Packaging Science
Contact
Associate Professor D. Barnett
Department of Analytical Chemistry
University of New South Wales
Sydney, NSW 2052, Australia,

CHAPTER 22
D.C. Levy and B. Naidoo
Contact
Dr D.C. Levy
Dept of Electrical Engineering
University of Sydney
Sydney, NSW 2006, Australia

THE SENSES: MEETING BIOLOGICAL NEEDS

D.M. STODDART

INTRODUCTION

The brain is fashioned to process information which flows to it from the outside world via the senses: sight, touch, hearing, taste and smell. Humans became essentially what they are today in the late Pliocene and early Pleistocene, about 2 million years ago. By that time they would have been quite recognisable as humans; little anatomical or physiological evolution has occurred since (Noble and Davidson, 1996), though the behaviour of modern humans is, on the whole, quite different from that of our Pleistocene relatives. The sense organs, together with their related operational physiology, and neural signal process signalling systems, are evolved structures which were moulded by natural selection operating on environmental selective pressures. Thus, when we smell 'Chanel No.5', taste the aroma of freeze-dried coffee volatiles, or listen to a Bach cantata, we do so with Palaeolithic noses, taste buds, and ears, designed by nature to serve the needs of hunter-gatherers of long ago. Our ancestors spent almost all of the last two million years as hunter-gatherers; the agrarian and intensely social lifestyle which typifies modern humans, developed just a second or two away from midnight on the 24 hour clock of human evolution. Our structure, brains and instinct enabled survival in a Pleistocene world. How we perceive the modern world is largely a product of our evolutionary past (Cosmides *et al.*, 1992).

The human sense organs did not all evolve at once. The evolutionary line that eventually gave rise to *Homo sapiens* can be traced back from the primates, through many branching dendrites to the earliest of the backboned animals, and further back to the single-celled origin of life itself. Animals that are anatomically simple, necessarily have simple sense organs, and single-celled animals have organelles which may have specialised functions. In evolutionary terms the most ancient of the senses is that which perceives chemical molecules — the so-called common chemical sense. Place an amoeba, taken from the bottom of a pond, in a glass dish, and watch it move around at random. Introduce a drop of vinegar at one side of the dish and you will notice the amoeba move away from this influence, seeking water of a neutral pH. The membrane surrounding the cell can detect this unfavourable environment, and it transmits this knowledge to the control centre in the nucleus. The result is a change in behaviour, every bit as profound as that which occurs when you pull down the sun visor in your car only to reveal a huntsman spider!

Of the senses we possess today, those of smell and taste are evolutionally the most ancient. Our world, and particularly our intellectual world, is one of sights and sounds bringing a continuous torrent of stimulation to our brains. With stereoscopic colour vision and acute hearing, humans are well-adapted for life on the open plains, where danger can be spotted at a distance, or heard as a leaf rustles under a pad or hoof. To lose the sense of sight or hearing is a disaster which few can overcome even with the use of complex modern electronic devices, but the loss of the sense of taste or smell generally causes little more than mild discomfiture. Is there a hierarchy of the senses? Will the sense of smell become a sensory equivalent of the human coccyx — a vestigial reminder of our evolutionary past? In this chapter I shall show that the sense of smell occupies a fundamental and fascinating place in the human psyche, and is far from redundant. It is anatomically simple — far simpler than the complex eye — but our understanding of its role in human biology is still fragmentary. This chapter will examine that role, and place it in a biological context.

THE SENSE OF SMELL

In physical terms the sense of smell consists of two tufts of spaghetti-like cilia lying high in the nasal cavity, which are connected to a special olfactory lobe of the brain. Unlike in the eye or the ear, where the receptor cells lie deeply buried behind many protective structures, the olfactory receptor cells of humans protrude into the outside world protected only by a thin layer of mucus. They are forward projections of the brain itself, and possibly the most exposed nervous tissue in the whole body. I shall leave a detailed description of the physiology and biochemistry of how the nose works to others, but suffice it to say that when odorant molecules are swept into the nasal cavity, they stimulate receptor sites on the spaghetti-like cilia. A wave of depolarisation flows down the axon of the receptor cell to the olfactory lobe of the brain. Just two or three synapses later and the signal, partially processed by various relay centres along the way, enters the so-called limbic system of the brain. In evolutionary terms the limbic system is the most ancient part of the brain; it is only in the higher mammals that the neocortex, or cerebral hemispheres, dominate. In fishes and other lower vertebrates the limbic system is known as the 'rhinencephalon' (or 'smell brain'), for so much of its activity is associated with the analysis of chemical molecules in the environment. In humans, the limbic system is thought to be the seat of emotion, with its many components interacting to control mood and temperament (Shepherd, 1983).

It is also the place where sexual behaviour and reproductive control is effected. Linked to the pituitary gland, or hypophysis, by a series of fine channels through which messenger substances can flow, the limbic system controls the pituitary gland and its production of the primary sex hormones: follicle stimulating hormone, and luteinising hormone. These two hormones stimulate the ovaries and the testes to mature and to produce their own sex hormones, oestrogen and testosterone.

Much is known of the role of smell in mammalian reproduction, and the interested reader is referred to some broad reviews for a comprehensive overview (eg Bronson, 1979; Stoddart, 1990). Odours are involved at almost all stages in mammalian reproductive biology: from initial attraction of the sexes and mate choice, to induction of oestrus, maintenance or termination of the pregnancy, appropriate nest-building, maternal-neonate imprinting, psychosexual development, and weaning. The scientific literature is richly endowed with the results of studies which show unequivocally the absolute requirement of a functional sense of smell for normal sexual reproduction to occur.

The situation in humans is far from clear, despite quite strenuous attempts to demonstrate that natural human odours play a role in human sexual behaviour and reproduction. There seems little doubt, however, that damage to certain parts of the olfactory system results in sexual dysfunction. Klingmueller and his colleagues have shown clearly that patients suffering from Kallmann's Syndrome, where the gonads fail to develop and sexual maturity never occurs, is accompanied by a severe deformation of a part of the brain known as the olfactory sulcus — a cleft in the olfactory part of the brain (Klingmueller et al., 1987). Although the physiological and biochemical pathways remain indeterminate as yet, it appears as if interruption of the normal neural pathways into the limbic system has an effect on the production of the chemicals which stimulate the pituitary into action. Whether human pheromones (natural aphrodisiac odours) exist, depends upon how critically one wishes to interrogate the evidence. In my view, the case for their existence has not been made beyond doubt, though there are many interesting and well-controlled studies which strongly suggest that we are more like other animals than we may care to admit. I await with great interest the developments of the next few years in this fascinating field.

THE SENSE OF SMELL IN HUMAN CULTURE

There are very few examples in the ethnographic literature of humans having developed an intellectual smell culture, but plenty indicating visual, acoustic, and gustatory intellectual culture. The ancient Japanese Tea Ceremony might be the best known exception. In his book *The World of the Shining Prince*, Ivan Morris (1964) describes how the blending of incense was an aristocratic art. During competitions the best judges rated the various concoctions for their different qualities, before pronouncing a winner. The Japanese language has adjectives that are used to describe different scents which cannot be adequately translated into English. 'Namamekashi', for instance, includes the suggestion of warmth, depth, damp; roughly translated as 'a deep, moist type of elegance'. Tea Ceremonies have all but disappeared in modern Japan, and they are, and were, unparalleled in the West. Perhaps it is because it projects to the seat of emotion, the sense of smell cannot be utilised for intellectual pursuits. The founding father of olfactory physiology, the Dutch neurophysiologist Hendrick Zwaardemaker, noted:

> We live in a world of odour as we live in a world of light and sound. But smell yields us no distinct ideas grouped in regular order, still less are they fixed in the memory as a grammatical discipline. Olfactory sensations make vague and half-understood perceptions, which are accompanied by very strong emotion. The emotion dominates us, but the sensation which was the cause of it remains unperceived. (Zwaardemaker, 1895)

The English essayist and psychologist, Havelock Ellis, drew a clear distinction between the sense of smell and the other senses — between the 'sense of imagination' and the 'senses of intellect', as he put it (Ellis, 1910).

> Sight is our most intellectual sense, and we trust ourselves to it with comparative boldness without any undue dread that its messages will hurt us by their personal intimacy; we even court its experiences, for it is the chief organ of our curiosity, as smell is of a dog's. But smell with us has ceased to be a leading channel of intellectual curiosity. Personal odours do not, as vision does, give us information that is very largely intellectual; they make an appeal that is mainly of an intimate, emotional, imaginate character. (Ellis, 1910)

Against this background, it is clear that the sense of smell is not like the intellectual senses, and being associated with emotion there is opportunity for appeals to be made to it in the manufacturing world.

The manufacture of perfumes goes back to the earliest of recorded history, and its use touches upon medicine, mythology, religion and anthropology. Perhaps the first recorded perfume was myrrh (the resin of *Commiphora myrrha*, a small Middle-Eastern shrub), recorded on an Egyptian papyrus from 4,000 BP (years before present). The Ancient Egyptians were of the view that, to ascend to the afterlife, a deceased person had to be purified with sweet-smelling unguents and incense such that the ferryman on the Styx river would let the spirit pass. Before being laid to rest in an appropriate place, the body was brought to Abydos to pay a personal visit to Osiris. An attendant priest constantly censed the body with myrrh and frankincense. Osiris was said to exude a fine odour to all who breathed his breath, and those who paid him a posthumous visit could be laid to rest in peace because they had breathed 'a breath of myrrh and incense' (Erman, 1894). Ancient mythology holds that Osiris was originally a sacred cedar tree imported into Egypt from the Lebanon, and in further recognition of his powers, Egyptian tombs and sarcophagi were often lined with cedar panels (Frazer, 1923).

The part played by incense and fine odours in the lives of Ancient

Egyptians cannot be underestimated. The hieroglyphic determinative for 'happiness' depicts a nose, so closely was the perception of fragrance identified with well-being. The ingredients were extremely costly, and centuries later, when the Roman Empire had expanded to its maximum extent and had enthusiastically embraced the incense culture, the pressure on the Treasury to pay for the imports rose to unsustainable heights. Some historians believe this pressure was a contributory factor bringing about the final collapse of the Empire. Some two centuries earlier the censors Publicus Lucinius Crassus

Plate I
Burning fragrant perfumes in Ancient Egypt. (LaGallienne, 1928)

and Lucius Julius Caesar had issued a proclamation forbidding the sale of 'foreign essences', but it did little good. When Emperor Nero buried his wife Poppaea, an amount of incense equal to a year's production from the whole of Arabia was burned.

The Ancient Hebrews apparently made no use of incense. The first Biblical record of its use is to be found in the book of Jeremiah.

> To what purpose cometh there to me incense from Sheba, and the sweet cane from a far country? (*Jeremiah 6:20*)

In the centuries leading up to Christ's birth, incense rituals became increasingly complex, reaching their peak of complexity in the temple of Herod, some 20 years before Christ. In the early days of Christianity incense was shunned and treated with scorn and revulsion, on account of the heavy use made of it by the Jews. When, in the fourth century AD, Constantine the Great inaugurated the Peace of the Church, all opposition to the use of incense in Christian worship fell away, and today the incense rituals in the Christian Church are every bit as complex as those which occurred in the temple of Herod.

THE POWER OF PERFUME

Despite the fact that we all live, to a greater or lesser extent, in an intellectual world where our cerebral hemispheres dominate and rationalise our lives, we are enormously concerned with odour. Many commonplace household materials are made fragrant by the addition of perfume. Our own bodies, so well-endowed with scent-producing glands, are every day scrubbed, scraped, lathered with fragrant soaps, depilated and coated with artificial, ie non-human, odours. The most expensive perfumes contain the anal gland secretions of beavers and civet cats, and preputial gland secretions of musk deer. Beavers, civet cats and musk deer use these materials in their sexual behaviour for mate attraction and territorial demarcation. How odd it is that humans, the most highly scented of all the apes, should deny the expression of such a fundamental aspect of their humanity! How strange, too, that they should agree, even desire, to smell of rutting deer or civet cats, along with floral fragrances! The individual body odour of other people remains a last great social taboo, though the body odour of a lover or a beloved one can feed many artistic muses. The anonymous writer of the Song of Solomon, and the 16th century English poet Robert Herrick are amongst many writers who have written about the body odour of a lover with exquisite tenderness, charm and personal vulnerability.

I have argued elsewhere, and at considerable length, that the human requirement not to smell of humans had its origins in the Pliocene, when the world's climate dried up somewhat, causing the break up of forests and the emergence of grasslands (Stoddart, 1986; 1990). The argument is complex and cannot be developed here, but it has its base in the genetic imperative for male and female to remain in a bonded partnership while the dependent offspring are absorbing the cultures of hunting and gregariousness. When the forests broke up and grasslands developed, there evolved a new source of animal protein — large terrestrial mammals. To a single human ancestor, armed with no more than sticks and stones, these creatures were unattainable. But to groups of individuals, which had the intellect to communicate with one another and to plan how to lay ambush on large ungulates, they represented a vast source of food. A necessary concomitant was gregariousness, a situation in which many bonded pairs and their offspring lived together. It should be noted that, zoologically speaking, carnivores are not gregarious (with a tiny number of exceptions). The ancestors of modern humans were a marked exception.

Gregariousness brings its own sociobiological and genetic problems. The breeding cycle of ancestral humans was probably much as it is now. At periodic intervals an egg would be shed from the ovary. In the vast majority of mammals this event is heralded widely, with visual, acoustic and olfactory displays. This is what happens in our anthropoid and primate relatives. The course of our evolution has seen ovulation become concealed. Menstruation, on the other hand, cannot be concealed, but the chances of fertilisation occurring during menstruation are small. I have proposed that, at some time in our past when our ancestors started living in large social groups, our sense of smell became desensitised to human sexual odours in order that the sociobiological and genetic demands of a monogamous (or sequentially monogamous) mating system may be met. When this desensitisation was incorporated into the genome, only memory traces of the sexual allure of natural odours remained. Perfumes containing sex attractants of other species of mammals stimulate those memory traces, without releasing the behaviours which were once associated with them, and which we see all about us in the animal kingdom. Some years ago I drew attention to the fact that resins and oils present in incense ingredients — phytosterols — were chemically related to animal steroids (Stoddart, 1985). I argued that the attention-seeking and holding effect of incense might be related to this similarity, since so many animal sex attractant pheromones contain steroid compounds.

CONCLUSION

In this short essay I have attempted to provide a biological and evolutionary context within which the senses, and particularly that of smell, can be set. An understanding of how they evolved their particular capabilities helps to establish the framework within which

Plate 2
Massaging of oils into the hair after bathing in Roman baths.
(LaGallienne, 1928)

manufacturers of foodstuffs and articles should approach the redesign of their products. In the end, all manufactured products, whatever their nature, must be perceived by the purchaser's senses. The sense of smell is particularly enigmatic, and I have tried to indicate, very briefly, why this should be so. Nevertheless, with its direct and powerful appeal to the emotions, it offers a rich field for innovative design and manufacture, and even for the sensory refurbishment of sick buildings. Its power should never be underestimated.

ACKNOWLEDGMENTS

Illustrations depicting burning of fragrant perfumes in Egyptian urns, and massaging of oils into hair after bathing in Roman baths, are from LaGallienne, 1928. Plate 3 is reproduced by permission of Syndics of Cambridge University Library.

REFERENCES

Bronson, F.H. (1979) The reproductive ecology of the house mouse. *Q. Rev. Biol.*, 54, 265–299.

Cosmides, L., Tooby, J. and Barkow, J.H. (1992) Evolutionary psychology and conceptual integration. In Barkow, J.H., Cosmides L., and Tooby, J. (eds), *The Adapted Mind*. Oxford University Press, Oxford.

Ellis, H. (1910) *Studies on the Psychology of Sex*. Random House, NY.

Erman, A. (1894) *Life in Ancient Egypt*. Tirad, H.M. (translator), Macmillan, London.

Frazer, J.G. (1923) *The Golden Bough; a Study in Magic and Religion*. Macmillan, London.

The Holy Bible, King James Version, Jeremiah 6:20.

Klingmueller, D., Dewes, W., Krahe, T., Brecht, G. and Schweibert, H.U. (1987) Magnetic resonance imaging of the brain in patients with insomnia and hypothalamic hypogonadism (Kallmann's Syndrome). *J. Clin. Endocrinol. Metab.*, 65, 581–584.

LaGallienne, R. (1928) *The Romance of Perfume*. Richard Hudnut, NY.

Morris, I. (1964) *The World of the Shining Prince; Court Life in Ancient Japan*. Oxford University Press, Oxford.

Noble, W. and Davidson, I. (1996) *Human Evolution, Language and Mind*. Cambridge University Press, Cambridge.

Shepherd, G.M. (1983) *Neurobiology*. Oxford University Press, Oxford.

Stoddart, D.M. (1985) Is incense a pheromone? *Interdisc. Sci. Rev.*, 10, 237–247.

Stoddart, D.M. (1986) The role of olfaction in the evolution of human sexual biology: an hypothesis. *Man*, 21, 514–520.

Stoddart, D.M. (1990) *The Scented Ape; the Biology and Culture of Human Odour*. Cambridge University Press, Cambridge.

Zwaardemaker, H. (1895) *Die Physiologie des Geruchs*. Engelmann, Leipzig.

Plate 3
Stripping Cassia
bark to make
incense.
(Syndics of
Cambridge
University
Library)

FROM SWEETS TO HOT PEPPERS: GENETIC VARIATION IN TASTE, ORAL PAIN, AND ORAL TOUCH

L.M. Bartoshuk, K.E. Cunningham,
G.M. Dabrila, V.B. Duffy, L. Etter, K.R. Fast,
L.A. Lucchina, J.M. Prutkin, D.J. Snyder

INTRODUCTION

For years, the method for assessment of taste blindness was a threshold procedure (Harris and Kalmus, 1949). Thresholds for PTC (phenylthiocarbamide) and other bitter compounds (including 6-*n*-propylthiouracil, ie, PROP) containing the $N - C = S$ group produce a characteristic bimodal distribution (tasters and nontasters). Today we use PROP to assess genetic taste variation because it lacks the sulfurous odour of PTC. In addition, we now use suprathreshold methods to characterise the range of this variation since thresholds describe only the bottom of the range, often failing to reflect perceptions of more intense stimuli.

In our early experiments, we showed that the differences between tasters and nontasters extended to compounds that did not contain the $N - C = S$ group. For example, tasters perceived the greatest bitterness from caffeine (Hall *et al.*, 1975) and the greatest sweetness from sucrose (Gent and Bartoshuk, 1983). These data proved incorrect the original view that nontasters simply lacked a receptor for the N − C = S group.

Discovery of supertasters

Our more recent work was prompted by the variability among tasters. We suspected that tasters could be subdivided into medium tasters (PROP is moderately bitter) and supertasters (PROP is extremely bitter) (eg, see Bartoshuk *et al.*, 1994). Accurate psychophysical measurement of the size of the differences between nontasters, medium tasters, and supertasters depends on methodological issues discussed below.

PSYCHOPHYSICAL METHODS TO CHARACTERISE SUPERTASTERS

Measurement of perceived intensity

In many psychophysical experiments, the aim is to compare sensations generated by different stimuli. This is a relatively easy task. Asking a subject to compare sensations separated by long time intervals is more difficult, but still logically possible because a single subject experiences both sensations. But what can we do when we need to compare sensations across subjects or groups (eg, nontasters and supertasters)? We cannot make absolute comparisons of perceived intensity across subjects. However, we can compare the ability of subjects to make relative comparisons.

Magnitude estimation

One of the most commonly used scaling methods is magnitude estimation (Stevens, 1969). This method produces ratio data; that is, a rating of '20' reflects a sensation that is twice as intense as a sensation rated '10'. In early studies, the experimenter set the rating for a designated standard. Systematic differences in perceived intensity of the standard across subjects could not be detected with this procedure.

In later studies, the standard was dropped and subjects were allowed to use any numbers they deemed appropriate. The experimenter could then 'normalise' or 'standardise' the data by selecting the standard, and multiplying each subject's data by a factor chosen to make that standard equal for all subjects.

Magnitude matching

The key to making comparisons across subjects is to have a standard that does not vary with the sensation of interest. The method that makes use of such a standard is called magnitude matching (Marks *et al.*, 1988). This approach is based on insights from cross-modality matching studies (eg, Stevens and Marks, 1965; 1980) which showed that we are able to make reliable matches of perceived intensity across sensory continua. In our early work, we used NaCl as our standard because, even if NaCl does not taste exactly the same to everyone, we believed that any variation would not be related to the taste of PROP. However, using *sound* as our standard (where subjects were asked to equate perceived saltiness to the intensity of an auditory signal), we found that NaCl tastes saltier to supertasters than to medium and nontasters (Bartoshuk *et al.*, 1998b). This means that earlier studies using NaCl as a standard underestimated the size of differences due to PROP status.

Despite this, NaCl, although conservative, still can serve as a useful standard for testing substances that show greater variation with PROP than NaCl itself. In addition, a PROP ratio (the bitterness of PROP divided by the saltiness of NaCl) works well for classifying subjects, because the ratio varies considerably from nontasters to supertasters (Bartoshuk *et al.*, 1994). NaCl is also more practical for testing subject groups.

The Green scale

Green and his students have developed a semantically labelled magnitude scale (see Figure 2.1) to be used in place of magnitude estimation (Green *et al.*, 1993; Green *et al.*, 1996). However, it is questionable whether adjectives serve the same function as the tone standards in magnitude matching. Not all subjects use adjectives like 'strong' to refer to the same perceived intensities. Still, many subjects appear to use adjectives in a relatively similar manner. Thus, if the sample size is large enough, raw Green Scale ratings show many of the same PROP effects as magnitude matching. To improve comparisons across subjects, the Green scale can be combined with magnitude matching (Snyder *et al.*, 1996).

Context can produce false supertaster effects

Rankin and Marks (1991) have shown that intense sensations of one taste quality can cause subsequently experienced sensations of a different taste quality to seem more intense. If subjects taste concentrated PROP, supertasters will have the most intense experiences of bitterness; this can cause subsequent stimuli to taste too strong. Obviously, this would lead to untrue associations between PROP tasting and other sensations. Consequently, it is essential that PROP be tasted after other stimuli or on separate days.

14

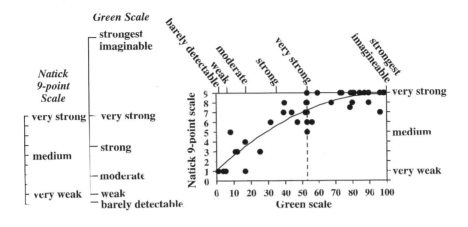

Figure 2.1

Bitterness of PROP paper measured with two methods: Natick 9-point scale and Green scale. The polynomial regression produces r = 0.84, p < 0.0001. The scales are shown on the left. Note that on the left side of the dotted line (marking 'very strong') the two scales show strong agreement. To the right of the dotted line, the 9-point scale fails to distinguish among subjects who give differing ratings with the Green scale.

Psychophysical errors can conceal supertaster effects

The error induced by designated standards (see discussion of magnitude estimation) can occur with any type of scale. For example, in a study using a line labelled 'no taste at all' on one end and 'extremely strong taste' at the other end, both nontasters and tasters of PROP were instructed to consider a given concentration of quinine to be 'maximum bitterness'. Needless to say, nontasters and tasters appeared to taste quinine as equally bitter under these conditions. In fact, with such a procedure, any stimuli that show PROP effects smaller than that shown by the standard can erroneously appear to taste more intense to nontasters than to tasters (eg, KC1 in Schifferstein and Frijters, 1991).

Psychophysical scales with ceiling effects may fail to separate medium tasters from supertasters. For example, attendees at a lecture on taste (N = 49) were asked to rate the maximum bitterness of PROP paper (about 1.6 mg PROP) with both the Natick 9-point scale and the Green scale. Approximately half were asked to use the Natick scale first, and half were asked to use the Green scale first. Approximately 45 minutes separated the two tests (see Figure 2.1).

The Green scale shows that supertasters rate PROP paper as more intense than 'very strong' when given the freedom to do so. When constrained by the 9-point scale, supertasters place saturated PROP close to the top ('very strong'). Upon scaling other tastants, super-tasters reduce their ratings of the weaker solutions proportionately. The result is that medium and supertasters produce very similar responses with the 9-point scale. In retrospect, we can see ceiling effects in several studies (Drewnowski, *et al.*, 1997a,b,c; Tepper and Nurse, 1997). At best, scales with ceiling effects make PROP effects look smaller than they really are. At worst, PROP effects may disappear entirely. For example, Lucchina *et al.* (1998) looked at the association of PROP bitterness with sucrose sweetness using both a 9-point category scale and the Green scale. The Green scale showed the expected association, but the 9-point category scale showed no association at all.

WHAT DO SUPERTASTERS EXPERIENCE?

Taste

Supertasters perceive greater intensities from most bitter substances and many (but not all) sweeteners (Bartoshuk *et al.*, 1992; Bartoshuk, 1993a,b). For example, supertasters experience much greater sweetness from sucrose and saccharin, but not from aspartame; thus there is no way to find concentrations of aspartame and sucrose that match for all individuals.

The graphs on the left in Figure 2.2 show recent data obtained with the Green scale (Prutkin, 1997). Subjects who had scaled PROP sipped and swallowed small amounts of sucrose and quinine on a later day. Note that supertasters (subjects with ratings at the high end of the Green scale) perceived the most intense sweetness and bitterness, but the effect for bitter was greater than that for sweet. This is typical.

ANATOMY: VIDEOMICROSCOPY OF THE TONGUE

Miller and Reedy (1990) found that blue food colouring does not stain fungiform papillae, the structures that contain taste buds; they can be seen as pink circles against a blue background. These investigators were the first to show differences in taste bud number between nontasters and tasters. In collaboration with them, we extended this work to supertasters and found that they had the most fungiform papillae.

To count fungiform papillae we (Reedy *et al.*, 1993; Bartoshuk *et al.*, 1994) asked subjects to insert their tongues between two plastic slides; and the fungiform papillae in a 3 mm square on the tip of the tongue were recorded on video (see Figure 2.2). Counts of papillae

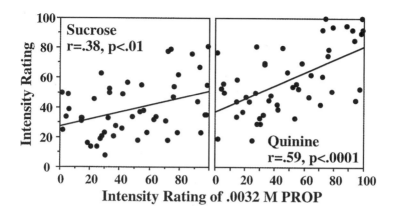

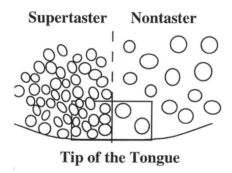

Tip of the Tongue

Figure 2.2
Graphs on the left: Perceived intensity of 1 M sucrose and 0.001 M quinine hydrochloride (sipped and swallowed) plotted as a function of the perceived intensity of 0.0032 M PROP (near saturated, not swallowed). Ratings were made with the Green scale. Note that the tastes of sucrose and quinine associate with PROP bitterness; supertasters perceive the most intense tastes. Drawing on the right: Sketch of the tongue tips of a supertaster and a nontaster.

may provide a more stable assessment of genetic status than PROP tasting because tasting reflects not only anatomy, but also taste changes resulting from various pathologies (eg, otitis media, head trauma) and hormonal cycling (see below).

Sex and race effects

Women are more likely than men to be supertasters (see Bartoshuk *et al.*, 1994). Early studies show racial variation in PTC tasting (Parr, 1934). A modern threshold study shows that Chinese have a low percentage of nontasters (Guo *et al.*, 1998) and our preliminary data suggest that Caucasians are more likely than Asian Americans to be nontasters.

Oral pain

Oral irritants commonly used in foods include capsaicin (chilli peppers), piperine (black pepper), ginger, and ethanol. Supertasters perceive the most intense oral burn from these irritants (Karrer and Bartoshuk, 1991; Karrer *et al.*, 1992; Snyder *et al.*, 1996; Prutkin *et al.*, 1998). Whitehead *et al.* (1985) showed that taste buds in fungiform papillae are surrounded by trigeminal neurons, which mediate pain and touch. To show that the PROP/pain association is due to anatomy, we applied 100 ppm capsaicin to tongue loci with and without taste buds. Supertasters rated burn as more intense on areas with taste buds, but not on areas without (Karrer *et al.*, 1992).

ORAL TOUCH: FAT

Fat molecules are too large to stimulate taste or smell, but do produce tactile sensations (eg, oily, or creamy) that are more intense to supertasters. This is because tactile sensations, like oral irritation, are mediated by the trigeminal nerve. Duffy *et al.* (1996) demonstrated this with a series of milk products varying in fat content. As fat content increases, supertasters perceive increasingly more 'creaminess' in the product. Salad dressings (Tepper and Nurse, 1997), corn oil, and guar gum (a thickening agent used in foods) (Prutkin, 1997; Prutkin *et al.*, 1999) also produce sensations that increase in perceived intensity with PROP taste.

Food preferences and nutritional consequences

The fact that supertasters show less liking for bitter foods is not surprising, but what are the consequences of heightened sensations for sweets and fats? In a sample of young adults, preferences for many high sweet and high fat foods varied with both PROP tasting and sex. In women, preferences for these same foods tended to fall as PROP tasting increased; for men, they tended to rise (Duffy *et al.*, 1995; Duffy and Bartoshuk, 1996).

Of special importance, food preferences were scaled using a method modified from Marks, *et al.* (1988). This method, devised for sensory scaling, consists of a horizontal line labelled 'nothing' on the extreme left and 'extremely strong' at a point two-thirds of the distance from 'nothing' to avoid ceiling effects. We modified this method for hedonic scaling by asking subjects to rate the intensity of their like or dislike on a horizontal line labelled 'neutral' on the extreme left and 'extremely' at the point two-thirds of the distance from 'neutral'. We believe that hedonic methods designed to avoid ceiling effects are likely to produce the best preference discriminations based on PROP status.

Do preference differences between males and females have a cultural or sensory origin? In American culture, females tend to worry about weight and this may lead to a dislike of the sensory properties of foods believed to be high in calories. On the other hand, PROP-associated distributions for females tend to be displaced, such that a subset of supertasting females are the most extreme supertasters. Perhaps the sensations evoked by sweet and fatty food that appeal to the average tongue are simply too intense for these individuals.

Two studies suggest a possible association between PROP-related food preferences and body weight. In elderly women, PROP tasting is associated with measures of body fat and serum lipids; women more responsive to PROP were thinner and had serum lipid values related to lower cardiovascular disease risk (Lucchina *et al.*, 1995). Younger supertasting women were also found to be thinner (Dabrila *et al.*, 1995).

Hormonal variation adds to genetic taste variation

Previous studies have suggested that the bitterness of PTC and PROP increases with pregnancy (eg, Bhatia and Puri, 1991). In a study (Duffy *et al.*, 1998) evaluating changes in bitter taste across pregnancy (N = 46), we found a significant rise in the bitterness of a moderate quinine concentration in the first trimester. Since many poisons are bitter, a rise in bitterness perception early in pregnancy would make women better poison detectors during critical times in foetal development.

Studies on the menstrual cycle have sought to link taste changes with phases of the cycle (eg, Beiguelman, 1964). Etter (unpublished thesis, 1999) asked subjects to taste PROP, NaCl, and sucrose upon awakening every day for three months. There was evidence of taste cycling, but the peak of the taste cycle did not always occur at the same phase of the menstrual cycle. This raises the possibility that a hormonal event acts as a trigger for a process that requires varying amounts of time across subjects. If this is true, taste ratings from cycling females should show large variance. We evaluated data sets from our laboratory and found support for this (Bartoshuk *et al.*, 1998a). For example, the bitterness of quinine showed greater variance in cycling women than in pregnant women. Also, the sweetness of sucrose and the creaminess of high fat dairy products showed greater variance for young women than young men. The burn of capsaicin (chilli peppers) showed greater variance for young than for postmenopausal women. One possible explanation for the fact that somatosensation and taste showed greater variance in women comes from a count of the number of taste buds per fungiform papilla. This number shows greater variance for young women than for young men. If the number of taste buds varies, then the trigeminal innervation associated with taste buds might be expected to vary as well.

CONCLUSION

Assessing PROP perception across individuals requires the most sophisticated psychophysical tools that we have. With the application of proper methodology, we find large differences in oral sensation that have important implications for food-related behaviour.

REFERENCES

Bartoshuk, L.M. (1993a) The biological basis of food perception and acceptance. *Food Qual. Pref.*, 4, 21–32.

Bartoshuk, L.M. (1993b) Genetic and pathological taste variation: What can we learn from animal models and human disease? In Bartoshuk, L.M. (ed), *Ciba Foundation Symposium 179*. John Wiley and Sons, NY.

Bartoshuk, L.M., Duffy, V.B., Etter, L., Fast, K., Garvin, V., Lucchina, L.A., Rodin, J., Snyder, D.J., Striegel-Moore, R. and Wolf, H. (1998a) Variability in taste, oral pain, and taste anatomy: Evidence for menstrual control over oral perception. *Appetite.*, 29, 388.

Bartoshuk, L.M., Duffy, V.B., Lucchina, L.A., Prutkin, J. and Fast, K. (1998b) PROP (6-n-propylthiouracil) supertasters and the saltiness of NaCl. In Murphy, C. (ed), *International Symposium on Olfaction and Taste XII*. NY Academy of Sciences, NY, 855, 792–796.

Bartoshuk, L.M., Duffy, V.B. and Miller, I.J. (1994) PTC/PROP tasting: Anatomy, psychophysics, and sex effects. *Physiol. Behav.*, 56, 1165–1171.

Bartoshuk, L.M., Fast, K., Karrer, T.A., Marino, S., Price, R.A. and Reed, D.A. (1992) PROP supertasters and the perception of sweetness and bitterness. *Chem. Senses*, 17, 594.

Beiguelman, B. (1964) Taste sensitivity to phenylthiourea and menstruation. *Acta Genet. Med. Gemellol.*, 13, 197–199.

Bhatia, S. and Puri, R. (1991) Taste sensitivity in pregnancy. *Indian J. Physiol. Pharmacol.*, 35, 121–124.

Dabrila, G.M., Bartoshuk, L.M. and Duffy, V.B. (1995) Preliminary findings of genetic taste status association with fat intake and body mass index in adults. *J. Am. Diet. Assoc.*, 95, A64.

Drewnowski, A., Henderson, S.A. and Shore, A.B. (1997a) Genetic sensitivity to 6-n-propylthiouracil (PROP) and hedonic responses to bitter and sweet tastes. *Chem. Senses*, 22, 27–37.

Drewnowski, A., Henderson, S.A. and Shore, A.B. (1997b) Taste responses to naringin, a flavonoid, and the acceptance of grapefruit juice are related to genetic sensitivity to 6-n-propylthiouracil. *Am. J. Clin. Nutr.*, 66, 391–397.

Drewnowski, A., Henderson, S.A., Shore, A.B. and Barratt-Fornell, A. (1997c) Nontasters, tasters, and supertasters of 6-n-propylthiouracil (PROP) and hedonic response to sweet. *Physiol. Behav.*, 62, 649–655.

Duffy, V.B. and Bartoshuk, L.M. (1996) Genetic taste perception and food preferences. *Food Qual. Pref.*, 7, 309.

Duffy, V.B., Bartoshuk, L.M., Lucchina, L.A., Snyder, D.J. and Tym, A. (1996) Supertasters of PROP (6-n-propylthiouracil) rate the highest creaminess to high-fat milk products. *Chem. Senses*, 21, 598.

Duffy, V.B., Bartoshuk, L.M., Striegel-Moore, R. and Rodin, J. (1998) Taste changes across pregnancy. In Murphy, C. (ed), *International Symposium on Olfaction and Taste XII,* NY Academy of Sciences, NY, 855, 805–810.

Duffy, V.B., Weingarten, H.P. and Bartoshuk, L.M. (1995) Preference for sweet in young adults associated with PROP (6-n-propylthiouracil) genetic taster status and sex. *Chem. Senses*, 20, 688.

Etter, L. (1999) Variations in taste perception across the menstrual cycle. Unpublished thesis for the MD degree, Yale University School of Medicine, New Haven, CT.

Gent, J.F. and Bartoshuk, L.M. (1983) Sweetness of sucrose, neohesperidin dihydrochalcone, and saccharin is related to genetic ability to taste the bitter substance 6-n-propylthiouracil. *Chem. Senses*, 7, 265–272.

Green, B.G., Dalton, P., Cowart, B., Rankin, K. and Higgins, J. (1996) Evaluating the labelled magnitude scale for measuring sensations of taste and smell. *Chem. Senses*, 21, 323–334.

Green, B.G., Shaffer, G.S. and Gilmore, M.M. (1993) A semantically-labelled magnitude scale of oral sensation with apparent ratio properties. *Chem. Senses*, 18, 683–702.

Guo, S.W., Shen, F.M., Zheng, C.J. and Wang, Y. (1998) Threshold distributions of phenylthiocarbamide (PTC) in the Chinese Population. In Murphy, C. (ed), *International Symposium on Olfaction and Taste XII,* NY Academy of Sciences, NY, 855.

Hall, M.J., Bartoshuk, L.M., Cain, W.S. and Stevens, J.C. (1975) PTC taste blindness and the taste of caffeine. *Nature*, 253, 442–443.

Harris, H. and Kalmus, H. (1949) The measurement of taste sensitivity to phenylthiourea (P.T.C.). *Ann. Eugenics*, 15, 24–31.

Karrer, T. and Bartoshuk, L. (1991) Capsaicin desensitization and recovery on the human tongue. *Physiol. Behav.*, 49, 757–764.

Karrer, T., Bartoshuk, L.M., Conner, E., Fehrenbaker, S., Grubin, D. and Snow, D. (1992) PROP status and its relationship to the perceived burn intensity of capsaicin at different tongue loci. *Chem. Senses*, 17, 649.

Lucchina, L., Bartoshuk, L.M., Duffy, V.B., Marks, L.E. and Ferris, A.M. (1995) 6-n-propylthiouracil perception affects nutritional status of independent-living older females. *Chem. Senses*, 20, 735.

Lucchina, L.A., Curtis, O.F., Putnam, P., Drewnowski, A. and Bartoshuk, L.M. (1998) Psychophysical measurement of 6-n-propylthiouracil (PROP) taste perception. In Murphy, C. (ed), *International Symposium on Olfaction and Taste XII,* NY Academy of Sciences, NY, 855, 816–819.

Marks, L.E., Stevens, J.C., Bartoshuk, L.M., Gent, J.G., Rifkin, B. and Stone, V.K. (1988) Magnitude matching: The measurement of taste and smell. *Chem. Senses*, 13, 63–87.

Miller, I.J. and Reedy, F.E. (1990) Variations in human taste bud density and taste intensity perception. *Physiol. Behav.*, 47, 1213–1219.

Parr, L.W. (1934) Taste blindness and race. *J. Heredity*, 25, 187–190.

Prutkin, J.M. (1997) PROP tasting and chemesthesis. Unpublished Senior Essay, Yale University.

Prutkin, J.M., Fast, K., Lucchina, L.A., and Bartoshuk, L.M. (1999) PROP (6-*n*-propylthiouracil) genetics and trigeminal innervation of fungifrom papillae. *Chem. Senses*, in press.

Rankin, K.M. and Marks, L.E. (1991) Differential context effects in taste perception. *Chem. Senses*, 16, 617–629.

Reedy, F.E., Bartoshuk, L.M., Miller, I.J., Duffy, V.B., Lucchina, L. and Yanagisawa, K. (1993) Relationships among papillae, taste pores, and 6-*n*-propylthiouracil (PROP) suprathreshold taste sensitivity. *Chem. Senses*, 18, 618–619.

Schifferstein, H.N.J. and Frijters, J.E.R. (1991) The perception of the taste of KCl, NaCl, and quinineHCl is not related to PROP-sensitivity. *Chem. Senses*, 16, 303–317.

Snyder, D.J., Lucchina, L.A., Duffy, V.B. and Bartoshuk, L.M. (1996) Magnitude matching adds power to the labelled magnitude scale. *Chem. Senses*, 21, 673.

Stevens, S.S. (1969) Sensory scales of taste intensity. *Perception Psychophys.*, 6, 302–308.

Stevens, J.C. and Marks, L.E. (1965) Cross-modality matching of brightness and loudness. *Proc. Natl. Acad. Sci. USA*, 54, 407–411.

Stevens, J.C. and Marks, L.E. (1980) Cross-modality matching functions generated by magnitude estimation. *Perception Psychophys.*, 27, 379–389.

Tepper, B.J. and Nurse, R.J. (1997) Fat perception is related to PROP taster status. *Physiol. Behav.*, 61, 949–954.

Whitehead, M.C., Beeman, C.S. and Kinsella, B.A. (1985) Distribution of taste and general sensory nerve endings in fungiform papillae of the hamster. *Am. J. Anat.*, 173, 185–201.

INTRODUCTION TO THE SENSE OF SMELL: UNDERSTANDING ODOURS FROM THE STUDY OF HUMAN AND ANIMAL BEHAVIOUR

S.L. YOUNGENTOB

INTRODUCTION

The olfactory system, whose millions of receptors are located high in the nasal cavity, is an ancient sensory system capable of detecting and discriminating among thousands of different odorants. Olfaction is absolutely critical to the survival of lower animals. For example, olfactory stimuli regulate reproductive physiology, food intake and social behaviour. In humans, the sense of smell is generally considered less critical to survival than the other special senses. However, the potential importance of this sense is being given new consideration today because of its enormous impact on the quality of life (consider the money spent in the perfume and food industry to modify attractiveness) and the possibility that, like in lower animals, chemical messengers may play an important role in the social and sexual behaviour of humans. In addition, disorders of the sense of smell can be profoundly distressing, as well as harbingers of more general disease states.

Given the continually emerging importance of this special sense (both physiologic and economic), the aim of this chapter is to briefly: 1) examine the important role that olfaction plays in human and animal behaviour; 2) examine the analytic problems faced by the olfactory system; and 3) review some research directions that have been taken, in an effort to understand how the olfactory system encodes information.

PERSPECTIVE

Olfaction is the sensory process that, in response to chemical stimuli, gives rise to the sensation called odour. The primary olfactory receptive area is that region of the nasal cavity subserved by the olfactory nerve. In humans, the olfactory receptors lie deep within the nasal cavity and are confined to a patch of specialised epithelium, the olfactory epithelium, covering roughly 5 cm^2 of the dorsal posterior recess of the nasal cavity (Moran et al., 1982). The olfactory receptor is a bipolar neuron that has a short peripheral process and a long central process.

The short peripheral process extends to the surface of the epithelium (which is in contact with the air space of the nasal cavity), where it ends in an expanded olfactory knob. This knob, in turn, gives rise to several cilia that, along with the cilia from other receptor cells, form a dense mat at the epithelial surface (Moran et al., 1982; Morrison and Costanzo, 1990). Odour transduction is initiated in these cilia as a result of the interaction of odorant molecules with specialised receptor proteins, within the ciliary membrane (Buck and Axel, 1991).

The longer central process of the olfactory receptor is an unmyelinated axon that projects through the cribriform plate to synapse on secondary projection neurons (mitral and tufted cells) in the olfactory bulb. Following interaction with local circuits within the olfactory bulb, the mitral and tufted cell axons project to higher cortical regions, including the piriform cortex (Shepherd and Greer, 1990).

Although the chemoreceptive endings and neural projections of the olfactory nerve are basic to the sensing of odours, other cranial nerves are involved — namely the trigeminal, glossopharyngeal and vagus. These accessory cranial nerves, which innervate different regions of the respiratory tract (the nose, pharynx and larynx), give rise to the pungent or irritating quality often experienced as part of an odour sensation (Parker, 1912; Cain, 1976; Cain, 1990). In addition, they also mediate a variety of reflexes in response to chemical stimulation (James and Daley, 1969). These reflexes tend to minimise the effects of noxious stimuli and protect the animal from further exposure. Among these reflexes are a decrease in respiratory rate, an increase in epinephrine secretion, an increase in nasal secretion, an increase in nasal air flow resistance, decreased heart rate, peripheral vasoconstriction, closure of the glottis, sneezing, and closure of the nares (James and Daley, 1969; Alarie, 1973; Eccles, 1990).

For much of the animal kingdom olfaction is basic to the maintenance of life. Olfactory stimuli regulate reproductive physiology, food intake, and social behaviour as well. For example, it is essential for finding prey and it serves as a first line of defence from becoming prey. It is used in communication: the male moth uses olfactory cues to find his mate 2.5 miles away (Schneider, 1969), as does the adult salmon to return to the place where it was spawned (Harden-Jones, 1968). Species as diverse as cats, dogs and deer mark their territory with urine or other secretions (Mech and Peters, 1977; Muller-Schwartz, 1977; Yahr, 1977), and when the skin of certain fish is damaged it releases substances to alarm nearby conspecifics (Smith, 1977).

Studies of a variety of species have shown a dependency of sexual behaviour upon olfactory cues. For example, male hamsters display mating, even with other males, when presented with vaginal discharge from receptive females (Darby, Devor and Chorover, 1975). The importance of this chemical communication should not be understated. In some animals, sexual dysfunction, and even retarded development of the sex organs results when the olfactory process is compromised (Brown, 1979; Gubernick, 1981; Hudsen and Distel, 1983; Doty, 1986; Poindron et al., 1988). Although not as extensively documented, a relationship between olfaction and sex also seems likely in humans. For example, at least for some odorants, olfactory acuity in women seems better at ovulation than during menstruation (Schneider and Wolf, 1955; Henkin, 1974), and there is strong evidence that olfactory cues (ie, human pheromones) among women, can synchronise the menstrual cycle (McClintock, 1983; Stern and McClintock, 1998).

In humans, unlike in the lower animals, the sense of smell is generally considered less critical to survival, although there are times when the detection of odours such as smoke, gas or decaying food can prevent bodily harm. Instead, civilised society seems to emphasise the impressive hedonic effect of olfaction. As examples of the positive affect, people add seasoning to their foods, perfumes to their bodies and incense to their homes. The importance of the negative effect is demonstrated by the number of commercial products for use against objectionable odours. These preferences of course depend upon a number of variables, such as age, sex, socio-ethnic background and previous odour experience (Wysocki *et al.*, 1991).

One instance in which olfaction plays a major role is, perhaps surprisingly, in flavour perception and the recognition of tastes. Much of what people think they taste, they actually smell. In one study asking subjects to identify 21 common food substances placed on the tongue (Mozell *et al.*, 1969), there was a decrease from an average of 60% correct to 10% correct when the nasal cavity was made inaccessible to any vapours given off by the substance. Even coffee and chocolate, which were correctly identified by over 90% of the test subjects when the nose was accessible, were not identified correctly by any subject when the nose was inaccessible. Thus, at least for humans, olfaction appears to have a tremendous impact on the quality of life, and anything that interferes with appropriate functioning can be profoundly distressing. Again, consider what happens to your simple appreciation of food when a cold or flu strikes.

PROBLEM FOR ANALYSIS

The olfactory system is a molecular detector of remarkable sensitivity. It has the capacity to discriminate among literally millions of odorants, including those that have never before been experienced. Similarly, the capacity to identify odorants on re-exposure is limited only by the paucity of vocabulary that can be used to describe them. Understanding the means by which the olfactory system encodes and decodes information has not been an easy task, given the lack of a clear physical energy continuum to characterise and control stimulus presentation. In other words, unlike in vision and audition there is no simple metric analogous to wavelength for colour or frequency for pitch. The situation is made even more complex by the finding that similar chemical substances have quite different odours. For example, d-carvone smells like caraway, and l-carvone smells like spearmint. Both are stereoisomers; their formulas are the same, but the two molecules are mirror images. On the other hand, some substances with quite different chemical formulas smell alike; for example carborane, trisallylrhodium and cyclopentadienyl-tricarbon monoxide-manganese all smell like camphor. So how does the olfactory system handle the encoding of odorant information?

Chemists have sought to understand the coding of olfactory information by comparing the structure of chemical compounds that smell 'alike' (as determined psychophysically). The hope was that after ordering all odorants (or some reasonable number) into as few groups as possible based on the perceptual similarity of their odours, one would be able to specify the physical or chemical similarity of the substances in each class. That similarity, in turn, would provide a clue to the nature of the stimulus, and then one would be in a position to study the neurophysiological transduction mechanisms.

Amoore proposed a stereochemical classification model for describing odour quality and similarity (Amoore, 1952, 1962a, 1962b). In this model, odorants with similar odours also have similar molecular sizes and shapes. For example, camphoraceous odours would all have molecules shaped like a ball, and there would be a bowl-like receptor site to accommodate it. In contrast, an ethereal odour would be shaped more like a sausage, and would need an oblong, dish-like receptor site. Amoore first proposed to explain all odour qualities on the basis of a set of seven primary odours or odour dimensions: ethereal, camphoraceous, musky, floral, minty, pungent and putrid. The concept was that the sense of smell operated on the basis of a limited number of discrete 'primary odour' sensations, which can be combined in different proportions to give a tremendous range of distinguishable odours. As such, smell would be analogous to taste with its four classic primaries of salt, sour, bitter and sweet, or to colour with its three primaries of red, yellow and blue. This would be in contrast to audition, which depends on a continuum of frequencies, with no primary notes.

The best evidence for primary odours was provided by a detailed study of smell blindness, or specific anosmias (ie, reduced sensitivity or insensitivity to one or a very limited number of odours in the presence of otherwise normal olfactory functioning) (Amoore, 1967, 1970). Amoore believed this phenomenon was genetic in origin and involved primary odours and specific receptors. Subsequent research has, in fact, suggested that more classes of odour primaries are required to account for all the different odour sensations experienced, and that the number of heritable anosmias are greater than seven. So Amoore's original notion of seven primaries was wrong, even though the concept may be correct. Amoore just didn't go far enough.

Molecular biological techniques have recently been used to identify a large family of receptor proteins in the olfactory epithelium, probably numbering somewhere in the order of 700–1000 (Buck and Axel, 1991; see also Sullivan, this volume). Therefore, there may be, in fact, 700–1000 odour primaries, each corresponding to a particular receptor protein. Exactly why nature may have adopted such a complex system is unclear. It may be that the necessary high degree of sensitivity can only be achieved when the olfactory receptor site is very closely matched to the shape and affinity of the odorant. In order to detect a wide range of compounds with the necessary high sensitivity, a large number of primary odour receptor types would be necessary.

In contrast to chemists, neurobiologists have attempted to understand olfactory coding by application of stimuli and examining the physiological and behavioural response. For example, there has been increasing evidence in both mammals and amphibians that the olfactory epithelium is excited by odorants in a highly distributed fashion, and that different odorants produce different spatial patterns of neural activity (Adrian, 1950, 1951; Mozell, 1966, 1970; Kauer and Moulton, 1974; Moulton, 1976; Thommesen and Doving, 1977; Kubie et al., 1980; MacKay-Sim and Kubie, 1981; MacKay-Sim et al. 1982; MacKay-Sim and Shaman, 1984; Mozell et al., 1987; Kent and Mozell, 1992; MacKay-Sim and Kesteven, 1994; Youngentob et al., 1995; Youngentob and Kent, 1995). In exploring the possibility that non-homogeneous patterns of sensitivity to different odorants exist along the olfactory mucosa, one approach has emphasised optical recordings from the epithelium of rats previously stained with a voltage-sensitive dye (Youngentob et al., 1995; Youngentob and Kent, 1995; Kent et al., 1995; Kent et al., 1996). The general strategy in these studies was to simultaneously monitor many mucosal sites (usually 100) in an optical matrix, and record the fluorescence changes in response to puffs of different odorants directed in a uniform fashion onto the tissue. A similar study, using these same techniques, has also been done for two amphibian species (Kent and Mozell, 1992). All these studies observed that different odorants gave their maximal responses at different mucosal regions, even though they gave at least some minimal response in all regions. These activity patterns, which reflect the differential odorant responsiveness of the receptor cell in different regions, were termed 'inherent' activity patterns by Moulton (1976) and his co-workers, who initially observed them using EOG (electro-olfactogram) recordings from the olfactory mucosa of salamander.

The exact underlying mechanism by which these spatial patterns are established is unknown. However, the simplest explanation is that receptor cells are distributed non-uniformly within the olfactory mucosa, such that those with similar responsiveness are located near each other in the epithelium. Evidence along these lines has emerged with the identification of topographically distinct patterns of receptor expression (Dear et al., 1991; Nef et al., 1992; Strotmann et al., 1992; Ressler et al., 1993; Vassar et al., 1993; Kubick et al., 1997; see also Sullivan, this volume). Although the broad expression zones are organised in bands across the mucosa, unlike the spatial patterns of neuronal activity observed in neurophysiological studies, some families and subfamilies of receptors show a clustered expression pattern (Strotmann *et al.*, 1992; Kubick, *et al.*, 1997). The expression area of these related receptor subtypes apparently fit the size of the regions of maximal activity, or 'hot spots', previously demonstrated using neurophysiolgical techniques (MacKay-Sim and Kesteven, 1994; Youngentob *et al.*, 1995; Kent *et al.*, 1996). Therefore, the segregation of these types of receptors in small regions of the olfactory mucosa could reflect the basis for the 'inherent' activity patterns initially described by Moulton (1976). In addition, the uniform spatial organisation of some receptor types does not obviate the finding that an odorant will activate, to different degrees, different receptor proteins.

Several investigators have shown that receptor cells are broadly tuned (Gesteland *et al.*, 1963; Getchell and Shepherd, 1978; Revial *et al.*, 1978, 1982, 1983; Firestein *et al.*, 1993). Thus, the subset of neurons expressing one receptor protein is likely to respond to some of the same odorants as other neurons expressing a different receptor, which lie in other distinct spatial zones. In short, one could imagine therefore how variations in the responsivity of gene family and subfamily members could establish the differential responsiveness observed with neurophysiological techniques. The formation of odorant-induced spatial activity patterns may occur through the grouping of olfactory neurons that express the same, or related, odorant receptors which have identical, or at least very similar, odorant specificities.

With regard to the above, one important question has been whether a relationship exists between a candidate coding mechanism, such as spatial activity patterns, and the perception of the animal. To explore this hypothesis, Kent *et al.* (1995) examined whether a predictive relationship existed between the relative position of an odorant's mucosal loci of maximal activity, and the relative position of the same odorant in a psychophysically determined perceptual odour space. To accomplish this, rats were trained to differentially report (ie, identify) the odorants propanol, carvone, citral, propyl acetate and ethylacetoacetate. Once trained, the animals were tested multiple times, using a confusion matrix design, and the resultant data was used to measure the degree of perceptual dissimilarity between any pair of the five odorants (Table 3.1). The dissimilarity measures were, in turn, subjected to multidimensional scaling analysis in order to establish a two-dimensional perceptual odour space for each rat (Figure 3.1).

Table 3.1

Composite behavioural data for a single animal from 40 testing sessions

		Stimulus Response Alternatives				
		PR	CA	CI	EA	PA
	PR	0.950	0	0.03	0.013	0.008
	CA	0	0.999	0.001	0	0
Stimulus	CI	0.019	0.005	0.938	0.028	0.011
	EA	0.011	0	0.013	0.975	0.001
	PA	0.008	0	0.015	0.003	0.975

$P(C) = 97\%$, $N = 4000$ trials (800) per odorant

For each particular odorant, the entries in the cells of that row of the matrix were the relative frequencies with which the animal responded with the correct response and the four incorrect alternatives (EA: ethylacetoacetate, PA: propyl acetate, CA: l–carvone, PR: propanol, and CI:citral). A comparison of the distributions of responses in any two rows (ie, odorants) of the composite matrix was performed by calculating the information transmitted (Atteneuve, 1959). By doing so, a measure of the degree of perceptual dissimilarity between any pair of odorants was obtained. (Adapted from Kent *et al.*, 1995)

A voltage-sensitive dye was applied to the nasal septum and to the medial surface of the turbinate for each behaviourally trained rat. At the completion of behavioural testing, the fluorescence changes in the dye were monitored, in response to the same odorants. Kent *et al.* found, as in previous studies, that although the entire monitored mucosal area responded to the presentation of an odorant, each of the five odorants used in the behavioural testing produced a distinct spatial distribution pattern of neural activity. More importantly, each odorant had a unique region of maximum sensitivity or 'hot spot' on the mucosa, that was consistent from animal to animal (Figure 3.2). Analysis of the behavioural and neurophysiological data indicated a highly significant predictive relationship between the relative position of an odorant's mucosal loci of maximal activity, and the relative position of the same odorant in a psychophysically determined perceptual odour space (Figure 3.3). In other words, if the area uniquely activated by one odorant was relatively farther apart on the mucosa from the area uniquely activated by another, then these same odorants would also be relatively distant from each other in the perceptual odour space.

Conversely, if the unique areas of differential mucosal activity were relatively closer together, then these same odorants would be relatively closer together in the odour space. Thus, on the basis of these data, it was suggested that the mucosal representation of an

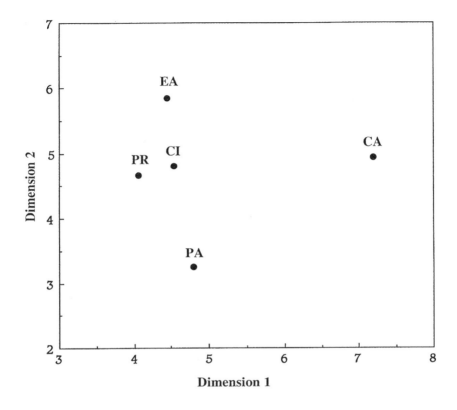

Figure 3.1

An illustration of the perceptual odour space determined from the composite data of Table 3.1, using multidimensional scaling (MDS) techniques. Dimensions 1 and 2 are arbitrarily oriented coordinates. (EA: ethylacetoacetate, PA: propyl acetate, CA: l–carvone, PR: propanol, and CI: citral). (Kent *et al.*, 1995)

odorant is indeed distributed such that large-scale spatial patterns of activation serve as the basis of odour perceptual space. The results also suggested that the *relational* information encoded in the odorant-induced mucosal activity patterns was preserved through further neural processing.

These conclusions were consistent with several observations from other investigators which suggested that there may be an organised and stereotyped patterning of information that occurs at the level of the peripheral projection from the olfactory mucosa onto the bulb. First, the anatomy of the olfactory system is well suited for the spatial segregation of afferent inputs emerging from different regions of the olfactory epithelium, at least to the level of the olfactory bulb

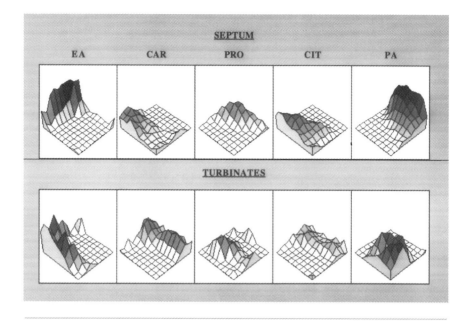

Figure 3.2
The surface plot data, which is the average across all animals in the study, emphasises the difference in the odorant-specific mucosal activity patterns for both the septum and turbinates. The magnitude of the difference of increased relative sensitivity for each pixel of the optical recording array is represented by the height on the z-axis and by 256 different shades of grey (the first level, white is assigned the value of no difference and black, the maximum value of 40% difference). (EA:ethylacetoacetate, PA: propyl acetate, CAR: l-carvone, PRO: propanol, and CIT: citral.) (Kent *et al.*, 1995)

(Shepherd, 1991). Second, and more importantly, the data on the regional segregation of odorant receptor expression in the mucosa (Ressler *et al.*, 1993; Vassar *et al.*, 1993; Kubick *et al.*, 1997), and their projection onto the olfactory bulb (Vassar *et al.*, 1994; Mombaerts *et al.*, 1996) suggests that each odorant will interact with a unique combination of receptors that are differentially distributed in epithelial space. Since any given odorant will interact with many different receptors, and a single receptor type will interact with many different odorants, stimulation with a single odorant would result in the activation of a unique combination of receptors, which in turn would result in the activation of a unique combination of secondary neurons in the olfactory bulb. Thus, it has been suggested that a fixed map of bulbar activation would permit the relational information laid down at the level of the mucosa to be transmitted to the brain (Kent *et al.* 1995).

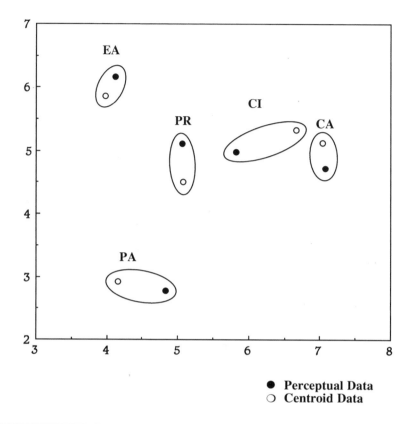

Figure 3.3
Alignment of the relative position of an odorant's mucosal loci of maximal activity and the relative position of the same odorant in a psychophysically determined perceptual odour space. Data are for the group average. The coordinates are arbitrary, and their scale was chosen in order to standardise the variance of the five locations (either behavioural or neurophysiological) to equal the number of dimensions, ie, two. (Kent *et al.*, 1995)

ACKNOWLEDGMENTS

The preparation of this article was supported by NIH-NIDCD grant.

REFERENCES

Adrian, E.D. (1950) Sensory discrimination with some recent evidence from the olfactory organ. *Br. Med. Bull.*, 6, 330–331.

Adrian, E.D. (1951) Olfactory discrimination. *Ann. Psychol.*, 50, 107–130.

Alarie, Y. (1973) Sensory irritation by airborne chemicals. *CRC Crit. Rev. Toxicol.*, 2, 299–363.

Amoore, J.E. (1952) The stereochemical specificities of human olfactory receptors. *Perfum. Essent. Oil Rec.*, 43, 321–330.

Amoore, J. E. (1962a) The stereochemical theory of olfaction. I. Identification of seven primary odors. *Proc. Sci. Sect. Toilet Goods Assoc.*, 37[suppl.], 1–12.

Amoore, J. E. (1962b) The stereochemical theory of olfaction. II. Elucidation of the stereochemical properties of the olfactory receptor sites. *Proc. Sci. Sect. Toilet Goods Assoc.*, 37[suppl.], 13–23.

Amoore, J.E. (1967) Specific anosmia: a clue to the olfactory code. *Nature*, 214, 1095–1098.

Amoore, J.E. (1970) *The molecular basis of odour*. Charles C. Thomas, Springfield, Il.

Attneuve, F. (1959) *Applications of Information Theory to Psychology*. Henry Holt and Co., NY.

Brown, R.E. (1979) Mammalian social odors: a critical review. *Adv. Stud. Behav.*, 10, 104–162.

Buck, L. and Axel, R. (1991) A novel multigene family may encode odorant receptors: a molecular basis for odour recognition. *Cell*, 65, 175-187.

Cain, W.S. (1976) Olfaction and the common chemical sense: some psychophysical contrasts. *Sensory Proc.*, 1, 57–67.

Cain, W.S. (1990) Perceptual characteristics of nasal irritation. In Green, B.G., Mason, J.R., Kare, M.R. (eds), *Chemical Senses*. Volume 2: Irritation. Marcel Dekker, NY, pp. 43–58.

Darby, E.M., Devor, M. and Chorover, S.L. (1975) A presumptive sex pheromone in the hamster: Some behavioural effects. *J. Comp. Physiol. Psychol.*, 88, 496–502.

Dear, T.N., Boehm, T., Keverne, E.B. and Rabbitts, T.H. (1991) Novel genes for potential ligand-binding proteins in subregions of the olfactory mucosa. *EMBO J.*, 10, 2813–2819.

Doty, R.L. (1986) Odour-guided behaviour in mammals. *Experentia*, 42, 257–271.

Eccles, R. (1990) Effects of menthol on nasal sensation of airflow. In Green, B.G., Mason, J.R., Kare, M.R. (eds), *Chemical Senses*. Volume 2: Irritation. Marcel Dekker, NY, pp. 275–295.

Firestein, S., Picco, C. and Menini, A. (1993) The relation between stimulus and response in olfactory receptor cells of the tiger salamander. *J. Physiol.(Lon.)*, 468, 1–10.

Gesteland, R.C, Lettvin, J.Y., Pitts, W.H., and Rojas, A. (1963) In Zotterman, Y. (ed), *Olfaction and Taste*. Pergamon Press, Oxford, pp. 19–44.

Getchell, T.V. and Shepherd, G.M. (1978) Responses of olfactory receptor cells to step pulses of odour at different concentrations in salamander. *J. Physiol.*, 282, 521–540.

Gubernick, D.J. (1981) Parent and infant attachment in mammals. In Gubernick, D.J. and Klopher, P.H. (eds), *Parental Care in Mammals*. Plenum Press, NY, pp. 243–305.

Harden-Jones, F.R. (1968) *Fish Migration*. Edward Arnold, London.

Henkin, R.I. (1974) Sensory changes during the menstrual cycle. In Ferin, M., Halberg, F., Richart, R.M. and Vande Wiele, R.L. (eds), *Biorhythms and Human Reproduction*. Wiley, NY, pp. 277–285.

Hudsen, R. and Distel, H. (1983) Nipple location by newborn rabbits: behavioural evidence for pheromonal guidance. *Behaviour*, 85, 260–275.

James, J.E.A. and Daley, M de B. (1969) Nasal reflexes. *Proc. Roy. Soc. Med.*, 62, 1287–1293.

Kauer, J.S. and Moulton, D.G. (1974) Responses of olfactory bulb neurones to odour stimulation of small nasal areas in the salamander. *J. Physiol.*, 243, 717–737.

Kent, P.F. and Mozell, M.M. (1992) The recording of odorant-induced mucosal activity patterns with a voltage-sensitive dye. *J. Neurophysiol.*, 68, 1804–1819.

Kent, P.F., Mozell, M.M., Murphy, S.J. and Hornung, D.E. (1996) The interaction of imposed and inherent olfactory mucosal activity patterns and their composite representation in a mammalian species using voltage-sensitive dyes. *J. Neurosci.*, 16, 345–353.

Kent, P.F., Youngentob, S.L. and Sheehe, P.R. (1995) Odorant-specific spatial patterns in mucosal activity predict perceptual differences among odorants. *J. Neurophysiol.*, 74, 1777–1781.

Kubick, S., Strotmann, J., Andreini, I., and Breer, H. (1997) Subfamily of olfactory receptors characterized by unique structural features and expression patterns. *J. Neurochem.*, 69, 465–475.

Kubie, J.L., MacKay-Sim, A. and Moulton, D.G. (1980) Inherent spatial patterning of response to odorants in the salamander olfactory epithelium. In Van der Starre, H. (ed), *Olfaction and Taste VII*. IRL Press, Oxford, pp. 163–166.

McClintock, M. (1983) Pheromonal regulation of the ovarian cycle: enhancement, suppression and synchrony. In Vandenbergh, J.G. (ed), *Pheromones and Reproduction in Mammals*. Academic Press, NY, pp. 95–112.

MacKay-Sim, A., and Kesteven, S. (1994) Topographic patterns of responsiveness to odorants in the rat olfactory epithelium. *J. Neurophysiol.*, 71, 150–160.

MacKay-Sim, A. and Kubie, J.L. (1981) The salamander nose: a model system for the study of spatial coding of olfactory quality. *Chem. Senses*, 6, 249–257.

MacKay-Sim, A., Shaman, P. and Moulton, D.G. (1982) Topographic coding of olfactory quality: odorant specific patterns of epithelial responsivity in the salamander. *J. Neurophysiol.*, 48, 584–596.

MacKay-Sim, A., and Shaman, P. (1984) Topographic coding of odorant quality is maintained at different concentrations in the salamander olfactory epithelium. *Brain Res.*, 297, 207–217.

Mech, L.D. and Peters, R.P. (1977) The study of chemical communication in free-ranging mammals. In Muller-Schwartz, D. and Mozell, M.M. (eds), *Chemical Signals in Vertebrates*. Plenum Press, NY, pp. 321–333.

Mombaerts, P., Wang, F., Dulac, C., Chao, S.K., Nemes, A., Mendelsohn, M., Edmondson, J. and Axel, R. (1996) Visualizing an olfactory sensory map. *Cell*, 87, 675–86.

Moran, D.T., Rowley, J.C. and Jafek, B.W. (1982) Electron microscopy of human olfactory epithelium reveals a new cell type: the microvillar cell. *Brain Res.*, 253, 39–46.

Morrison, E.E. and Costanzo, R.M. (1990) Morphology of the human olfactory epithelium. *J. Comp. Neurol.*, 297, 1–13.

Moulton, D.G. (1976) Spatial patterning response to odors in the peripheral olfactory system. *Physiol. Rev.*, 56, 578–593.

Mozell, M.M. (1966) The spatiotemporal analysis of odorants at the level of the olfactory receptor sheet. *J. Gen. Physiol.*, 50, 25–41.

Mozell, M.M. (1970) Evidence for a chromatographic model of olfaction. *J. Gen. Physiol.*, 56, 46–63.

Mozell, M.M., Sheehe, P.R., Hornung, D.E., Kent, P.F., Youngentob, S.L., and Murphy, S.J. (1987) 'Imposed' and 'inherent' mucosal activity patterns: their composite representation of olfactory stimuli. *J. Gen. Physiol.*, 90, 625–650.

Mozell, M.M., Smith, B.P., Smith, P.E., Sullivan, R.L. and Swender, P. (1969) Nasal chemoreception in flavor identification. *Archiv. Otolaryngol.*, 90, 131–137.

Muller-Schwartz, D. (1977) Complex mammalian behaviour and pheromone bioassay in the field. In Muller-Schwartz, D. and Mozell, M.M. (eds), *Chemical Signals in Vertebrates*. Plenum Press, NY, pp. 413–435.

Nef, P., Hermans-Borgmeyer, I., Artieres, H., Beasely, L., Dionne,V.E. and Heinemann, S.F. (1992) Spatial pattern of receptor gene expression in the olfactory epithelium. *Proc. Natl. Acad. Sci. USA*, 89, 8948-8952.

Parker, G.H. (1912) The reaction of smell, taste and the common chemical sense in vertebrates. *J. Acad. Nat. Sci. Phila.*, 15, 221–234.

Poindron, P., Levy, F. and Krehbiel, D. (1988) Genital, olfactory and endocrine interactions in the development of maternal behaviour in the parturient ewe. *Psychoneuroendocrinology*, 13, 99–125.

Ressler, K.J., Sullivan, S.L. and Buck, L.B. (1993) A zonal organization of odorant receptor gene expression in the olfactory epithelium. *Cell*, 73, 597–609.

Revial, M.F., Duchamp, A. and Holley, A. (1978) Odour discrimination by frog olfactory receptors: a second study. *Chem. Senses*, 3, 7–21.

Revial, M.F., Sicard, G., Duchamp, A. and Holley, A. (1982) New studies on odour discrimination in the frogs olfactory receptor cells. I. Experimental results. *Chem. Senses*, 7, 175–190.

Revial, M.F., Sicard, G., Duchamp, A. and Holley, A. (1983) New studies on odour discrimination in the frogs olfactory receptor cells. II. Mathematical analysis of electrophysiological responses. *Chem. Senses*, 8, 179–194.

Schneider, D. (1969) Insect olfaction: Deciphering system for chemical messages. *Science*, 163, 1031–1036.

Schneider, R.A. and Wolf, S. (1955) Olfactory perception thresholds for citral utilizing a new type olfactorium. *J. Appl. Physiol.*, 8, 337–342.

Shepherd, G.M. (1991) Computational structure of the olfactory system. In Davis, J.L. and Eichenbaum, H. (eds), *Olfaction: A Model System for Computational Neuroscience*. MIT Press, Cambridge, MA, pp. 3–42.

Shepherd, G.M. and Greer, C.A. (1990) The olfactory bulb. In Shepherd, G.M. (ed), *Synaptic organization of the brain*. Oxford University Press, NY, pp. 133–169.

Smith, R.J.F. (1977) Chemical communication as adaptation: Alarm substance of fish. In Muller-Schwartz, D. and Mozell, M.M. (eds), *Chemical Signals in Vertebrates*. Plenum Press, NY, pp. 303–321.

Stern, K. and McClintock, M.K. (1998) Regulation of ovulation by human pheromones. *Nature*, 392, 177–179.

Strotmann, J., Wanner, I., Krieger, J., Raming, K. and Breer, H. (1992) Expression of odorant receptors in spatially restricted subsets of chemosensory neurones. *Neuroreport*, 3, 1053-1056.

Thommesen, G., and Doving, K.B. (1977) Spatial distribution of EOG in the rat: a variation with odour quality. *Acta Physiol. Scand.*, 99, 270–280.

Vassar, R., Chao, S.K., Sitcheran, R., Nunez, J.M., Vosshall, L.B. and Axel, R. (1994) Topographic organization of sensory projections to the olfactory bulb. *Cell*, 79, 981–991.

Vassar, R., Ngai, J. and Axel, R. (1993) Spatial segregation of odorant receptor expression in the mammalian olfactory epithelium. *Cell*, 74, 309–318.

Wysocki, C.J., Pierce, J.D. and Gilbert, A.N. (1991) Geographic, cross-cultural and individual variation in human olfaction. In Getchell, T.V., Doty, R.L., Bartoshuk, L.M. and Snow, J.B. (eds), *Smell and Taste in Health and Disease*. Raven Press, NY, pp. 287–314.

Yahr, P. (1977) Central control of scent marking. In Muller-Schwartz, D. and Mozell, M.M. (eds), *Chemical Signals in Vertebrates*. Plenum Press, NY, pp. 549–563.

Youngentob, S.L., and Kent, P.F. (1995) Enhancement of odorant-induced mucosal activity patterns in rats trained on an odorant identification task. *Brain Res.*, 670, 82–88.

Youngentob, S.L., Kent, P.F., Sheehe, P.R., Schwob, J.E. and Tzoumaka, E. (1995) Mucosal inherent activity patterns in the rat: evidence from voltage-sensitive dyes. *J. Neurophysiol.*, 73, 387–398.

INTRODUCTION TO THE TRIGEMINAL SENSE: THE ROLE OF PUNGENCY IN FOOD FLAVOURS

J. PRESCOTT

INTRODUCTION

Sensations of pungency in food are usually considered as characteristic of non-Western cuisines, such as those from Mexico, India, China, or Korea — countries in which part of the defining 'flavour principle' (Rozin, 1985) includes ingredients such as chilli, pepper or ginger. However, the essential character of a wide variety of Western foods and beverages — carbonated drinks, wine, onion, mustard and horseradish, to give only a few examples — is also dependent on the presence of significant amounts of pungency. A cola drink without the fizz; the glass of wine without its sharpness; or onions, mustard and horseradish without their bite or ability to induce tears have lost much of their defining quality. The loss of this quality is primarily due to one factor — reduced pungency due to loss of stimulation of the trigeminal nerve in the nose or mouth.

Pungency represents an important, yet often overlooked, component of flavour. For example, consideration of pungency in discussions of the sensory basis of food acceptability is often limited to a fraction of the attention given to taste or odour. In fact, in many respects, pungency may be considered as essential as the sense of taste in our appreciation of many foods. Irritation produced by stimulation of the trigeminal nerve in the mouth or nose, which can be generally referred to as *pungency* when encountered in a food context, is produced by a wide variety of substances used as food ingredients, including chilli, pepper, mustard, ginger, menthol, carbon dioxide (CO_2) and alcohol (a more comprehensive list of irritants, together with their sources, is given in Table 4.1 on page 41). Sodium chloride, as well as many food acids such as citric and acetic acids, in high concentrations, have irritant as well as taste properties (Green and Gelhard, 1989; Gilmore and Green, 1993). Nasal pungency is also produced by odorous food components such as vinegar and mustard. In fact, whenever we describe a food as pungent, piquant, tangy, sharp, tingling, biting, spicy or, in the case of peppermint, cooling, we are typically referring to trigeminal stimulation.

Pungency as a food characteristic is ubiquitous throughout the world. If we consider only one irritant, chilli, it is estimated that more than 25% of the world's population, spread across Asia, South and Central America, Africa, and Europe, consume chilli every day (Rozin, 1990). Even more commonplace are the pungent qualities of the world's most utilised food ingredient, salt. Other pungent ingredients, eg vinegar, pepper, and ginger, are widely used throughout China and other parts of Asia. Even Japan, whose cuisine is generally low in pungency (at least compared to other Asian countries), uses *wasabi* (horseradish) noted for its high level of nasal and oral pungency.

Determining the role of pungency in foods is clearly essential to understanding the basis of food acceptability in Western cuisines, especially since, due to recent patterns of Asian immigration, Western diets are increasingly influenced by cuisines (such as those from Thailand and Vietnam) that use high levels of chilli and other pungent spices. In addition, research into pungency will provide insights into the food cultures of a variety of countries, especially in Asia. In turn, this will assist food companies in the production of more appropriate foods for export to Asia.

PUNGENT QUALITIES

What dimensions of sensory impact does pungency add to foods? In general terms, pungency may range from mild sensations of warmth or irritation, to sensations that are noxiously painful. Within food, pungency is typically described using a variety of poorly defined terms such as spicy, hot or piquant. While the traditional view of this sensory system is of a 'common chemical sense' that responds only to stimulus intensity, there is some evidence that different irritating substances can be distinguished to some extent by the particular combination of sensations that they produce. Qualitative differences have even been reported for the various pungent compounds within chillies (Todd *et al.*, 1977).

Table 4.1 shows some of the qualities reported to be associated with some common pungent compounds. The concentration of the pungent compound will determine to some extent the qualities perceived, the extent to which one or more qualities dominate, and also the duration of the sensation. Not only do irritants frequently produce more than one sensation, but these sensations may vary independently of one another (see, for example, Cliff and Green, 1994; Prescott and Stevenson, 1996a). The site of stimulation within the mouth is also important, as some areas may be more sensitive than others (Green and Lawless, 1991).

The pungency produced by a number of common irritants present in food ingredients (including capsaicin, cinnamaldehyde, eugenol, cuminaldehyde, piperine, ethanol and zingerone), have been shown to be characterised by their ability to produce sensations of burning, tingling, or numbing. The possibility of making clear distinctions between different pungent compounds on the basis of these qualities has been suggested. For example; capsaicin has been described primarily in terms of a burning sensation; alcohol, a bite; and menthol, both numbing and stinging. In addition, irritants can be distinguished by the lag to produce sensations, the duration of sensations, and the site within the mouth (tongue location, palate, etc.) where the sensations are maximal. Factors such as maximum intensity, rate of onset, and rate of decay have also been used to distinguish the sensations produced by cinnamaldehyde, capsaicin and piperine (Cliff and Heymann, 1992). Such findings have raised the question of whether or not the trigeminal system is capable of qualitative distinctions in a way analogous to the taste system's distinction between sweet, sour, salty and bitter.

Because of their taste properties, sodium chloride and food acids such as citric and acetic acid are seldom considered to be pungent compounds. However, both these compounds have been shown to also produce distinct irritation, often at concentrations used in foods. Sodium chloride, at levels that do not themselves produce

Table 4.1

Some common pungent food ingredients and their sources. Details are also given of the sensations that have been reported for these irritants.

Irritant	Source	Pungent sensations
allyl isothiocyanate	mustard	—
capsaicin	chilli	burning; numbing; tingling; stinging; biting
carbon dioxide	carbonated beverages	burning; stinging; tingling; biting; numbing
cinnamaldehyde	cinnamon	stinging/pricking; numbing; burning; warmth; tingling
citric/acetic acids	fruits/vinegar	—
cuminaldehyde	cumin	burning; tingling; numbing
eugenol	cloves	numbing
gingerol/zingerone	ginger	burning; numbing; tingling
piperine	pepper	stinging; burning; numbing; tingling; itching; pricking
2-propenyl /2-phenylethyl isothiocyanate	horseradish (wasabi)	—
menthol	peppermint oil	cooling; numbing; burning; tingling; stinging
sodium chloride	salt	numbing; stinging/pricking; burn

Sources: Cliff and Green (1994); Cliff and Heymann (1992); Cliff and Heymann (1993); Green (1988); Green (1991); Harper and McDaniel (1993); Lawless and Stevens (1988).

irritation, has also been shown to increase the burn of a mixture that contains capsaicin (Prescott *et al.*, 1993). It is likely, in fact, that a proportion of what we perceive as the taste impact of these substances is due to their pungency. This raises the possibility of using an irritant other than sodium chloride to compensate for reductions in salt, in those cases where salt-reduced foods are recommended for health reasons.

Studies suggest that the development of reliable and detailed characterisations of the pungent qualities of food ingredients could lead to a taxonomy of pungency, which could then be used to tailor the pungent characteristics of foods. However, as Table 4.1 shows, many diff⸗.ent pungent compounds share qualitatively similar sensations. Considerably more research is required that focuses on, firstly, developing reliable terms to describe pungency and, secondly, delineating the changes in quality produced by changes in concentration, method of stimulation, and spatio-temporal factors.

FACTORS INFLUENCING PUNGENCY

As with taste and odour, pungency is seldom experienced in isolation, and the context within which it is experienced influences both its intensity and quality. A variety of factors have been demonstrated to impact on the perception of pungency, among them temperature and other qualities which are present within foods. It is not surprising that the sensations produced by many trigeminal irritants are temperature dependent if we consider that the same sensory system is responsible for our experiences of heat and cold in the oral cavity. Taking the internal temperature of the mouth (37°C) as the baseline, research to date has shown that sensations of cooling and numbness are intensified by decreases in the temperature of solutions (Green, 1985); conversely, sensations of burn and warmth are intensified by temperature increases (Green, 1986a). In turn, pungent sensations can act to influence temperature perception (Green, 1986b). It is unclear whether or not the relationships of pungency and temperature found with simple solutions generalise to food.

The impact of tastes and other food qualities on pungency has received only limited attention. It has been shown that food with a significant fat content will reduce chilli burn (Baron and Penfield, 1996), most probably as a result of the high lipophilicity of capsaicin. However, merely chewing or introducing a solution into the mouth reduces capsaicin burn (Naswari and Pangborn, 1990a,b). The evidence that tastes (eg, sweetness) reduce pungency is also not reliable, although, as mentioned earlier, salt has been shown to enhance chilli burn, a finding understandable in terms of the pungent effects of sodium chloride on its own (Prescott et al., 1993).

CONTRASTS BETWEEN PUNGENCY AND TASTES

The dynamics and temporal properties of oral pungency show many differences from that of taste. Compared to tastes, pungency typically has a slow onset and persists, depending on strength, for prolonged periods — minutes to tens of minutes (Lawless, 1984). Thus, tastes are most intense for the few seconds following initial sampling, while they are in the mouth, whereas an irritant such as capsaicin has been shown to reach maximum intensity after tens of seconds, and following expectoration. Tastes tend to show adaptation, a reduction in intensity upon repeated sampling or continuous stimulation (Gent and McBurney, 1978). By contrast, multiple, successive samples of capsaicin produces sensitisation, or increases in intensity of the irritation (Green, 1989; 1991). However, this phenomenon is dependent on the time between successive samples. A sufficient pause in stimulation following a series of capsaicin solutions produces a marked decrease in sensation intensity for subsequent capsaicin solutions (desensitisation). The extent to which other irritants also produce sensitisation and desensitisation is not well established, although there is some doubt that sensitisation is a universal feature of oral irritation (Cliff and Green, 1994; Prescott and Stevenson, 1996b; Prescott, 1999).

Do sensitisation and desensitisation occur during food consumption? If they do occur, then the perception of irritant sensations within foods will change over time during their consumption. These effects may perhaps also influence the perception of other qualities, such as tastes. One possibility is that the sweetness of foods, for example, might decrease during consumption, through increasing inhibition during sensitisation, or increase through inhibition-release following desensitisation (see below). Other tastes (eg, saltiness), might show opposite effects, since their own irritant properties may act synergistically with capsaicin irritation. Clearly, if we wish to understand perception of the sensory qualities of foods with significant levels of oral pungency, it is important to establish

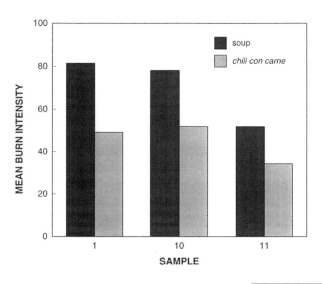

Figure 4.1
Mean burn intensity ratings during self-paced eating for the first, tenth and eleventh (post-hiatus) samples of soup and chili con carne, showing little change over the initial ten stimuli but evidence of desensitisation to the final stimulus.

whether sensitisation and desensitisation occur during food consumption. This question was recently examined in a study of two foods, a soup and a *chilli con carne* type dish, to which capsaicin had been added (Prescott, 1999). There was little evidence of sensitisation in either food, whether the consumption was self-paced or used the same timing that prior studies of capsaicin had used. Given that an associated study had shown sensitisation using capsaicin solutions, the failure to find sensitisation is most likely related to the complexity of the food matrix within which the capsaicin is embedded, in particular, its fat content. In contrast, desensitisation occurred when a hiatus was inserted following the initial series of samples, during the consumption of both foods and in both timing conditions (see Figure 4.1). This suggests that individuals may experience this effect following the consumption of spicy foods, although the magnitude appears to be much less than that found with capsaicin solutions. How long this desensitisation persists, and whether it is cumulative in frequent consumers of foods which produce high levels of burn, remains to be determined.

The quality and intensity of pungency produced by a variety of food ingredients is thus a complex function of concentration and time. These properties of pungency have important implications for product development, and implications for foods consumed with, or subsequent to, pungency. In the sensory evaluation of foods with significant pungency, the effects of sensitisation and desensitisation must be taken into account. For example, in an evaluation of multiple samples of foods with pungency, one sample may have a profound effect on the intensity of subsequent samples, depending on the extent to which the samples sensitise/desensitise, and the temporal spacing of the samples. Another implication of understanding the quantitative aspects of pungency is that it opens up the possibility that the quality of food flavour impact could be tailored using different pungent compounds. However, to date, there is little evidence regarding the effects of such manipulations.

Another possibility is the substitution of one pungent compound for another with similar stimulation characteristics. Unpublished observations from our laboratory suggest that, given an appropriate context (eg, the presence of orange flavor, sweetness and sourness), low levels of capsaicin irritation may be interpreted as carbonation. Again, this effect depends upon temporal factors, with the presence of a burning quality clearly distinguishing capsaicin from carbonation after 20 seconds or so. The utility of such findings may be in those cases where an irritant that is important for flavour impact is removed from a food or beverage for other reasons. A good example of this is low alcohol beer, which has poor consumer acceptability due to its lack of flavour impact.

INTERACTIONS BETWEEN PUNGENCY AND FOOD TASTES AND FLAVOURS

There is undoubtedly strong interaction between pungency and other sensory qualities: the total flavour impact of a variety of foods (including spicy foods, carbonated and alcoholic beverages) is dependent upon combinations of taste, flavours, and pungency. Despite this, relatively little is known about the ways in which pungency interacts with other components of food flavours to produce overall flavour impact. Cultural studies and much anecdotal evidence have suggested that for those who are very regular eaters of chilli, foods which lack pungency are bland and unappetising (Rozin, 1990). This may merely reflect a contrast effect between intensely pungent flavours and weaker flavours without pungency, although it has been shown that desensitisation produced by a large capsaicin dose produces decreases in taste intensity for one or more days afterward (Karrer and Bartoshuk, 1995). This raises the possibility that chronic desensitisation may produce chronic decrements in taste or flavour intensity. It might be expected that such effects would be most pronounced for salty and sour tastes, because of their concurrent pungency.

It has also been suggested that highly pungent food ingredients, such as chilli, may reduce the taste and flavour intensity of foods into which they are incorporated, especially for those unused to eating spicy foods. The evidence regarding tastes suggests that the only reliable effect is for the suppression of sweetness by capsaicin in both solutions and foods (Prescott *et al.*, 1993; Prescott and Stevenson, 1995). Irritation produced by CO_2 on the other hand, does not appear to suppress any tastes (Yau and McDaniel, 1992). Evidence to date suggests that flavour volatiles are not suppressed, even by quite strong levels of capsaicin pungency (Prescott and Stevenson, 1995).

PUNGENCY AS A FLAVOUR ENHANCER

Understanding the way in which trigeminal stimulants combine with tastes and flavours to produce a total flavour impact is important, given the ubiquity of such stimulants in foods. It is recognised that taste mixture components do not simply act additively in producing a total mixture intensity (McBride, 1989). Similarly, mixtures of oral irritants have been shown to have both additive and synergistic interactions, depending on the concentration (Lawless and Stevens, 1989). The interaction of tastants, flavours and irritants in determining total intensity, has received little attention, but is potentially very important to showing the relative role of each component in contributing to the flavour impact of foods.

Despite the fact that it may simply dominate mixtures if sufficiently intense, this does not preclude pungency acting in other ways within a food context. The role that pungency plays in real foods can be divided into two kinds. Firstly, pungency may increase total flavour impact without actually adding qualities such as burning. This will depend, to a large extent, on the intensity of pungency employed. In model solutions, it has been shown that adding citric acid to a sucrose solution may increase the overall taste intensity without increasing the sourness of the mixture, due to the suppression of sourness by sweetness (McBride and Finlay, 1990). Similarly, salt adds impact to bread flavour without making bread taste salty. Utilising pungency in this role may open up considerable opportunities for producing food flavours with enhanced impact.

One recent study investigated if oral irritation produced by capsaicin at low concentrations would enhance the flavour intensity of foods — in this case tomato and apple purées (Prescott and Francis, 1997). A group of 10 trained sensory panellists rated the flavour intensity of the purées to which had been added a range of capsaicin concentrations, starting at concentrations slightly above threshold. Subjects made continuous ratings until the flavour could no longer be perceived, but there was no evidence that low concentrations of capsaicin could enhance the flavour intensity of foods. In fact, quite the contrary occurred with the apple purée — the highest levels of capsaicin suppressed the flavour intensity, an effect most likely to be due to the suppression of the sweetness component of the flavour (see Figure 4.2).

Irrespective of its impact on flavour intensity, pungency may be able to add specific qualities, as outlined earlier, to food flavours. To date, the majority of research has been conducted using model systems, such as taste solutions. As such, there are few data dealing with the potential for pungency in this role. It is essential to examine the qualitative aspects of what pungency adds to food flavours within the context of both model and real foods. To what extent, for example, do the qualities (numbing, burning, stinging, etc.) possessed by different irritants carry over into real food flavours? How do these qualities alter the profile of these flavours? To what extent do these qualities impact on the consumer acceptance of products?

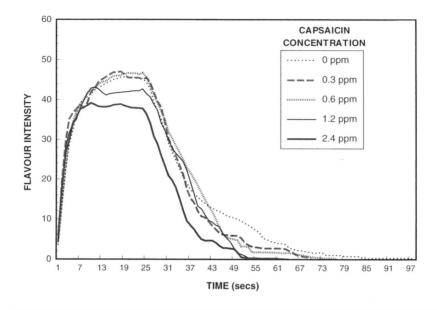

Figure 4.2

Time-intensity measures of apple purée to which had been added a range of capsaicin concentrations. There is no evidence that low concentrations of capsaicin increase the flavour intensity. The early decline in flavour intensity in the 2.4 ppm capsaicin condition is most likely due to sweetness suppression.

CONCLUSIONS

Research addressing these questions on the quantitative and qualitative effects of pungency within foods will shed light on the question of how liking for pungency develops and is maintained. It will also reveal the potential of using pungency to create novel food flavours with impact. For developers of food products for either domestic or export markets, understanding the influence of pungency on food perception and preference offers the opportunity of exploiting a sensory system that, to date, has been neglected by the food industry.

REFERENCES

Baron, R.F. and Penfield, M.P. (1996) Capsaicin heat intensity — concentration, carrier, fat level, and serving temperature effects. *J. Sens. Studies,* 11, 295–316.

Cliff, M.A. and Green, B.G. (1994) Sensory irritation and coolness produced by menthol: Evidence for selective desensitization of irritation. *Physiol. Behav.,* 56(5), 1021–1029.

Cliff, M.A. and Heymann, H. (1992) Descriptive analysis of oral pungency. *J. Sens. Studies,* 7, 279–290.

Cliff, M.A. and Heymann, H. (1993) Time-intensity evaluation of oral burn. *J. Sens. Studies*, 8, 201–211.

Gent, J.F. and McBurney, D.H. (1978) Time course of gustatory adaptation. *Perception Psychophys.*, 23(2), 171–175.

Gilmore, M.M. and Green, B.G. (1993) Sensory irritation and taste produced by NaCl and citric acid: effects of capsaicin desensitization. *Chem. Senses*, 18(3), 257–272.

Green, B.G. (1985) Menthol modulates oral sensations of warmth and cold. *Physiol. Behav.*, 35, 427–434.

Green, B.G. (1986a) Sensory interactions between capsaicin and temperature in the oral cavity. *Chem. Senses*, 11(3), 371–382.

Green, B.G. (1986b) Menthol inhibits the perception of warmth. *Physiol. Behav.*, 38, 833–838.

Green, B.G. (1988) Spatial and temporal factors in the perception of ethanol irritation on the tongue. *Perception Psychophys.*, 44(2), 108–116.

Green, B.G. (1989) Capsaicin sensitization and desensitization on the tongue produced by brief exposures to a low concentration. *Neurosci. Lett.*, 107, 173–178.

Green, B.G. (1991) Temporal characteristics of capsaicin sensitization and desensitization on the tongue. *Physiol. Behav.*, 49, 501–505.

Green, B.G. and Gelhard, B. (1989) Salt as an oral irritant. *Chem. Senses*, 14(2), 259–271.

Green, B.G. and Lawless, H.T. (1991) The psychophysics of somatosensory chemoreception in the nose and mouth. In Getchell, T.V., Bartoshuk, L.M., Doty, R.L. and Snow, J.B. (eds), *Smell and Taste in Health and Disease*. Raven Press, NY, pp. 235–253.

Harper, S.J. and McDaniel, M.R. (1993) Carbonated water lexicon: Temperature and CO_2 level influence on descriptive ratings. *J. Food Sci.*, 58(4), 893–898.

Karrer, T. and Bartoshuk, L. (1995). Effects of capsaicin desensitization on taste in humans. *Physiol. Behav.*, 57, 421–429.

Lawless, H.T. (1984) Oral chemical irritation: psychophysical properties. *Chem. Senses*, 9(2), 143–155.

Lawless, H.T. and Stevens, D.A. (1988) Responses by humans to oral chemical irritants as a function of locus of stimulation. *Perception Psychophys.*, 43, 73–78.

Lawless, H.T. and Stevens, D.A. (1989) Mixtures of oral chemical irritants. In Laing, D.G., Cain, W.S., McBride, R.L. and Ache, B.W. (eds), *Perception of Complex Smells and Tastes*. Academic Press, Sydney, pp. 297–309.

McBride, R.L. (1989) Three models for taste mixtures. In Laing, D.G., Cain, W.S., McBride, R.L. and Ache, B.W. (eds), *Perception of Complex Smells and Tastes*. Academic Press, Sydney, pp. 265–282.

McBride, R.L. and Finlay, D.C. (1990) Perceptual integration of tertiary taste mixtures. *Perception Psychophys.*, 48, 326–330.

Nasrawi, C.W. and Pangborn, R.M. (1990a) Temporal effectiveness of mouth-rinsing on capsaicin mouth-burn. *Physiol. Behav.*, 47, 617–623.

Nasrawi, C.W. and Pangborn, R.M.(1990b) Temporal gustatory and salivary responses to capsaicin upon repeated stimulation. *Physiol. Behav.*, 47, 611–615.

Prescott, J. (1999) The generalizability of capsaicin sensitization and desensitization. *Physiol. Behav.*

Prescott, J., Allen, S. and Stephens, L. (1993) Interactions between oral chemical irritation, taste and temperature. *Chem. Senses*, 18(4), 389–404.

Prescott, J. and Francis, J. (1997) The effects of capsaicin on flavor intensity in foods: A time intensity study. *Chem. Senses*, 22(6), 772.

Prescott, J. and Stevenson, R.J. (1995) The effects of oral chemical irritation on tastes and flavors in frequent and infrequent users of chili. *Physiol. Behav.*, 58(6), 1117–1127.

Prescott, J. and Stevenson, R.J. (1996a) Psychophysical responses to single and multiple presentations of the oral irritant zingerone: Relationship to frequency of chili consumption. *Physiol. Behav.*, 60(2), 617–624.

Prescott, J. and Stevenson, R.J. (1996b) Desensitization to oral zingerone irritation: Effects of stimulus parameters. *Physiol. Behav.*, 60(6), 1473–1480.

Prescott, J. and Swain-Campbell, N. (1998) *Responses to repeated oral irritation by capsaicin, cinnemaldehyde and ethanol in PROP tasters and non-tasters.* Paper presented at the European Chemosensory Research Organisation conference, Sienna, Italy, September 8–12th.

Rozin, E. (1985) Ethnic Cuisine: The Flavour-Principle Cookbook, The Stephen Greene Press, Battleboro, Vermont.

Rozin, P. (1990) Getting to like the burn of chili pepper. Biological, psychological and cultural perspectives. In Green, B.G., Mason, J.R. and Kare, M.R. (eds) *Chemical Senses, Volume 2: Irritation.* Marcel Dekker Inc, NY, pp. 231–269.

Todd, P.H., Bensinger, M.G., Biftu, T. (1977) Determination of pungency due to capsicum by gas-liquid chromatography. *J. Food Sci.*, 42(3), 660–665, 680.

Yau, N.J.N. and McDaniel, M.R. (1992) Carbonation interactions with sweetness and sourness. *J. Food Sci.*, 57(6), 1412–1416.

MEETING INDUSTRY NEEDS: THE EUROPEAN PERSPECTIVE

D.H. LYON

INTRODUCTION

The trend towards globalisation is having an effect on the food and drink industries, particularly in Europe. For example, Birds Eye Wall (a Unilever company) phased out all eight of its sub-brands in June 1996. Globalisation trends have placed even more pressure on sensory science to provide appropriate and reliable methodologies to guide the development of food products for the pan-European consumer market.

THE PAN-EUROPEAN MARKET

The European Union now embraces 15 countries (total population over 365 million), making a potential market more than thirty percent larger than that of the United States. Add to this the move towards harmonised laws and regulations, the dismantling of barriers to trade between partners, the prospect of economic monetary union, the enormous potential in the emerging consumer markets in Eastern Europe — and the opportunities for the food and drink industry become clear. However, companies are having to work harder in a much more complicated market, with a wide cultural, lingual and historical diversity. Only those companies that understand their consumers, and understand what influences consumer attitudes and drives preferences in food choice, will be successful in the long term.

We are already seeing evidence of this re-focus on consumer-led product development in the United Kingdom. The UK marketplace is probably the most dynamic in Europe, in terms of new food and drink product launches, stimulated by the strong influence of our major retailers. The Product Intelligence section of my own Department, which monitors all new food and drink product introductions in the UK market place, has seen a steady decline in numbers of new products launched in the last few years. I believe this reflects a sea change, with companies much more likely to target their new product development (NPD) to ensure that products launched in the market are more sure of success. This change of emphasis is reflected also by new proposals for research, requested and supported by CCFRA (Campden and Chorleywood Food Research Association) industrial members, into the role of sensory attributes in product success, and the establishment of methods to identify and measure causes of product success and failure.

EUROPEAN SENSORY NETWORK

So, how are those working in sensory science responding to the changes in the development and marketing of food and drink in Europe? In June of 1996, the European Community Research Programme re-launched an initiative to support the establishment of thematic networks. In the context of stimulating international trade, the European Commission sees these networks as a way of bringing together researchers from industry, universities, and research institutes, to co-ordinate industrially relevant research and development carried out separately in each country. The European Sensory Network (ESN) was established well ahead of this, in 1989, with exactly these aims. We now have 15 member organisations in 12 different countries in Europe, and one in Canada, and see this as the start of a truly world-wide sensory network for the future.

Members of the ESN are common in their aim of stimulating the development of new methods in applied sensory science, of standardising sensory methods, if appropriate, and of promoting the use of sensory analysis through food industry conferences and seminars.

An international training course is organised every year, pooling expertise from all members, to deliver the right sort of training at the right level to the food industry in different countries. Research workers are interchanged between members to stimulate new thinking, and to further stimulate co-operation and co-ordination. In addition, we encourage the setting up of 'national working groups', bringing industry together in a common forum to interact on a non-competitive level, sharing expertise and experience.

One of the most significant outputs of the UK working party based at CCFRA is the *Guidelines for Sensory Analysis in Product Development and Quality Control* (Lyon *et al.*, 1992). This is widely seen as unique in giving practical hands-on guidance to sensory analysts in industry. A new edition is due to be published in July 1999 (Aspen Publishers, USA).

From the beginning, however, even within the members of the ESN, it was clear that there were marked differences in the way each organisation approached such routine tasks as difference or descriptive testing using trained sensory panels. But did it matter? We quickly realised that if sensory science was going to help companies such as Birds Eye to develop pan-European products, then we needed to know the answer to this. Will Birds Eye get the same results if they profile their products at the innovation centre in the UK, as they would at other centres in the rest of Europe? Will they be able to communicate sensory results between centres in a common language? Members of the ESN have been addressing this issue on two broad fronts: accreditation, and standardisation, of sensory evaluation methodology and practices.

STANDARDISATION AND ACCREDITATION

One of the frequent comments I hear from both technical and marketing people in food and drink companies is that sensory analysis is 'soft science', and so subjective that the results are difficult to interpret and therefore of limited use. That is in spite of the fact that sensory descriptions are so basic to the definition and specification of products that companies cannot trade efficiently without detailed sensory specifications. Indeed, within the European Union, marketing regulations are in force for certain products, which specify their sensory characteristics in great detail (EC, 1991). I do not dispute that there are some bad sensory practices around which might justify these opinions in some instances, but I strongly defend sensory testing with trained panels as an objective and reliable analytical procedure.

Mutual recognition of analytical test methods and results between member states is seen by the European Union as a means of supporting regulations, which underpin the single European market and aid cross-border trade. The *Official Control of Foodstuffs Directive* (EC, 1993) requires that all analyses conducted under the provision of the directive be carried out by laboratories which follow a common quality standard, and are subject to a common inspection service. In summary, this means that laboratories are required to comply with the criteria set out in *EN45001: General criteria for the operation of testing laboratories* (CEN/CENELEC, 1989), and that recognised laboratories, responsible for assessing all laboratories, shall comply with the criteria set out in *ISO Guide 25: General requirements for the competence of calibration and testing laboratories* (ISO, 1990).

In practice, these measures have emphasised the need for analytical laboratories to obtain recognised accreditation, not least because having a laboratory's organisation, operations and test methods independently assessed and accredited helps manufacturers to demonstrate due care and attention under such legislation as the *Product Liability Directive* (EC, 1985). At CCFRA we have wholeheartedly endorsed the need for accreditation in all our laboratories, and for our organisation as a whole. Furthermore, we have campaigned hard to include sensory testing within the scope of our accreditation. We have worked hard to demonstrate that by paying due attention to traceability, reproducibility, calibration and auditing, sensory methods are as equally applicable to external accreditation as any other analytical method. In 1994, my department was the first laboratory in the UK to receive independent national accreditation for sensory methods. Furthermore, this pattern was repeated in other countries, as more and more sensory laboratories are now having their systems and organisation independently accredited.

In 1996, a milestone was achieved with the publication of the *Accreditation of Sensory Testing Laboratories* by the European Co-operation for Accreditation of Laboratories (EAL), (EAL, 1995). This document supplements EN45001 and ISO Guide 25, and provides specific guidance on the accreditation of sensory testing laboratories for both assessors and laboratories preparing for accreditation. It also gives detailed guidance on the interpretation of both documents, for those undertaking sensory analysis. We have made it: Sensory scientists now have their methods on a level playing field. So let's get on with winning the game!

Members of the ESN have been very keen to answer two key

questions: Do sensory profiles from one panel differ from sensory profiles of the same product from another panel in another country, or indeed in the same country? If so, why and how do they differ? Secondly, can sensory profiles from a panel in one country be used to relate to the preference or acceptability of the product as perceived by consumers in another country? Clearly, sensory methods must be able to help determine which sensory attributes of products are important in their appeal to consumers, if manufacturers are to use these methods to help create new products for different countries. Furthermore, the least companies expect is that sensory scientists are consistent amongst themselves, and that the methods they use will provide similar results on identical products, irrespective of which analyst or panel is used.

In order to increase our understanding of these issues, ESN members undertook, and recently reported on, a large inter-laboratory study on roast and ground coffee (ESN, 1996). The study concerned the sensory descriptive analysis of sixteen different coffees, representing a range typical of those consumed throughout Europe, followed by consumer testing of eight of the coffees selected from this range.

The descriptive analysis was carried out by 11 members of the ESN in eight different European countries. We standardised everything possible: the coffee samples, method of preparation, temperature of serving, experimental design and method of presentation. We even standardised the water used to make the coffee, and shipped bottled mineral water from the same source throughout Europe. What we did not standardise, however, was the way in which the test was carried out. Here we had laboratories, each with recognised sensory expertise, and some with formal accreditation. We wanted to know whether it mattered that we had different approaches to selection and training of panels, that we had different ways of generating descriptors, or that we used different numbers of attributes. If it did matter, the results would show that we needed to work harder towards standardisation of approach, and application of recognised methodology.

The results of this trial were astonishing. We do not know of any other trial that has been carried out with this degree of control of experimental details, but no standardisation of the panel approach to the work. Furthermore, we included extremes in terms of panel expertise. On the one hand were the highly professional descriptive panel of the International Coffee Organisation, with vast experience in coffee tasting. On the other hand were panels which had been recruited specifically for this work, with no experience of coffee tasting. We worked in eight different languages with terms ranging in number from 14 to 56 attributes. Panel training time also varied from 8 hours to up to 96 hours. However, what we found when we analysed the results using multivariate statistics was very clear. Each panel generated data to yield a very similar sample structure; in other words, each panel had perceived the differences between the samples in a similar way, but had used different words and languages to describe those differences. The main difference between the panels was one of degree of separation of samples, with the more 'expert' panels more confident at separating the samples. What is more, the separation of the samples by each panel made sense, considering what was known about the coffee samples in the coffee trade. (Of course this information was not communicated to the panels prior to testing.)

The study was also designed to bring together the attributes from the different panels and, in so doing, to propose a common vocabulary that could be understood by all panels. This could directly reflect the situation in companies where, for example, a common vocabulary is necessary so that it can be used by different panels on different sites. Detailed analysis of the data was carried out to establish whether a common meaning was a reasonable and justified assumption for bringing attributes together, or whether different panels placed different emphasis on different attributes. It was possible to propose a common vocabulary of 13 attributes: bitter, burnt, rubber, roast, chemical, strength of coffee flavour, acid, fruity, citrus, sour, rancid, astringent and floral (ESN, 1996).

A principal component analysis (PCA) was performed on the selected European vocabulary of 13 attributes, and the results were presented as a biplot (Figure 5.1). In terms of sample structure, this European biplot is visually similar to those obtained from individual panels, and the relationship between attributes also reflects those of individual panels.

For food and drink companies in Europe, however, understanding product differences is not enough. Companies really want to know what consumers like or dislike, how these consumers are grouped or segmented across Europe, and particularly, whether there are groups of consumers with similar likes and dislikes, and whether these can be explained in terms of sensory descriptions identified by the trained panel. ESN members followed up the sensory trial with eight separate consumer trials in seven different countries, using a sub-sample of the original coffee profiles by the sensory panels. Again, the same experimental procedures were followed by each organisation.

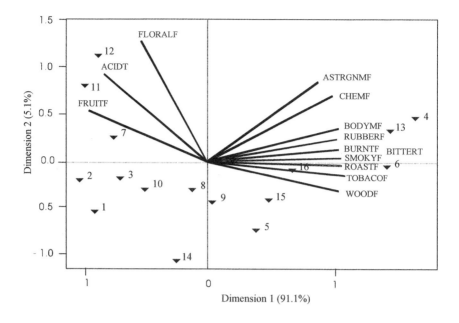

Figure 5.1

Sample and attribute biplot derived from covariance PCA on the European vocabulary. Samples are shown as 1-16. Attributes: FLORALF floral flavour; ACIDT acid taste; FRUITF fruit flavour; ASTRGNMF astringent mouth-feel; CHEMF chemical flavour; BODYMF body mouth-feel; RUBBERF rubber flavour; BURNTF burnt flavour; BITTERT bitter taste; SMOKYF smoky flavour; ROASTF roast flavour; TOBACOF tobacco flavour; WOODF wood flavour.

The data from all consumer trials were pooled and analysed by cluster analysis, as it was thought more interesting to segment consumers into groups with similar likes and dislikes, rather than by country. It was quite clear from the results that different clusters could be identified, and that the consumers making up each cluster were not all from the same country. We were more interested, however, in relating this consumer preference data to that obtained from sensory profiling, to identify directions of common preference and the attributes driving preference. The data from these clusters were then used to define a preference map onto which the European profile of 13 attributes were projected. In this way, we could understand what attributes were driving preferences for different groups of consumers (see Figure 5.2). Once again, such techniques have distinct advantages for industries which wish to develop and manufacture products in one country, confident that they will meet the requirements of the consumers in their target market.

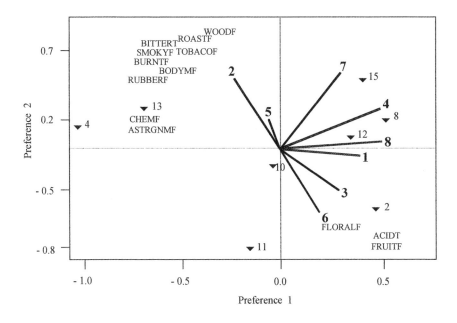

Figure 5.2
PCA biplot representing the sample positions in relation to the main preference directions.

PROFICIENCY TESTING SCHEMES

The results of these trials have given us new information, but where do they take us from here? Perhaps we could learn more by repeating this trial on other products in other circumstances and with other laboratories? This we aim to do, but the next key project for the ESN is to try to get some structure into inter-laboratory trials so that we can all learn from each other's results, and step by step take our discipline forward. At our meeting in October 1996, we agreed to champion the need for an International Harmonised Protocol for Proficiency Testing in Sensory Analysis, which sets down a common framework for inter-laboratory trials and proficiency testing schemes. Such a protocol will meet the need of accreditation authorities and help individual laboratories to fulfil their obligations to participate in inter-laboratory trials and to keep and maintain records of panel and assessor performance. This proposal has now been financially supported by the European Commission, and we hope to involve as many countries and organisations as possible to make this a success.

MEETING INDUSTRY NEEDS

So how are we in Europe using sensory science to meet industry needs? In summary, my team at CCFRA and my colleagues in the ESN are not content to see sensory science as a little-understood tool buried somewhere deep in the Research and Development department of food companies, if it exists at all. Understanding what drives consumer preferences, and using this information to help develop products which meet the needs of consumers, is the key to the future success of food companies in Europe. Industry needs practical tools on which it can rely. We are determined to demonstrate that sensory science offers those practical tools to face the challenges for marketing products in a wider European market.

REFERENCES

CEN/CENELEC (1989) EN45001. *General Criteria for the Operation of Testing Laboratories.*

EAL (1995) *EAL-G16:Accreditation for Sensory Testing Laboratories - Guidance on the Interpretation of the EN45000 Series of Standards and ISO/IEC Guide 25.*

EC (1985) Council Directive 85/374/EEC on the approximation of the laws, regulations and administrative procedures of Member States concerning liability for defective products. *Off. J. EC,* L210, 28, 7.8.85.

EC (1991) Commission Regulation (EEC) No 2568/91 of 11 July 1991 On the characteristics of olive oil and olive residue oil and on the relevant methods of analysis. *Off. J. EC,* L248, 34, 5.9.91.

EC (1993) Council Directive 93/99/EEC on the subject of additional measures concerning the official control of food stuffs. *Off. J. EC,* L290, 36, 24.11.93.

European Sensory Network (1996) *A European Sensory and Consumer Study — A Case Study on Coffee.* Available c/- CCFRA Chipping Campden, Glos., GL55 6LD UK.

ISO (1990) *Guide 25: General Requirements for the Competence of Calibration and Testing Laboratories.*

Lyon, D.H., Francombe, M.A., Hasdell, T.A. and Lawson, K. (1992) *Guidelines for Sensory Analysis in Food Product Development and Quality Control.* Chapman and Hall, London.

CHAPTER

6

J. WELLER

THE VALUE OF SENSORY SCIENCE TO THE FOOD PRODUCT DEVELOPER AND FOOD PROCESSOR

INTRODUCTION

New product introductions often fail as some literature suggests that as many as eight products will fail for every ten launched. This is a very costly process for companies. Uncle Tobys not only has a high rate of launching new and innovative products, but also manages to keep them on the shelf. A ninety percent product success rate is achieved by combining the talents of a multi-disciplinary team, and a structured, sophisticated project management approach.

Sensory data from consumers and trained panels have been integrated into the process, providing information for making objective decisions, and contributing to 'getting it right the first time'. The process is focused on understanding and interpreting the consumers' real wants. This reduces the overall cost of development, decreases the total development time, and increases the products' chances of success in the market place.

Sensory data is also being used during the manufacturing process for a number of reasons, including formula optimisation, ingredient change, specification evaluation, consumer complaint evaluation, and raw material taint testing.

THE VALUE OF SENSORY SCIENCE

'Value-adding', 'adding value', and 'value-adding steps' are phrases we often hear today. What comes to mind when you think about value? Your answer probably depends on which field of endeavour you are involved in. If you are an accountant or a manufacturing person, you probably think about the almighty dollar, or maybe magnitudes of a measurement or number. If you are involved in marketing or scientific research, you may think about intrinsic worth or importance.

As a food processor, Uncle Tobys' aim is to provide value to our consumers so that they will buy our products. The value gained from the product by the consumer varies with the consumer's wants and needs. The value that the consumer receives comes in many forms. Value for money is an important one, but it is not the only one. Nutritional value, a trusted brand, quality, consistency; and less tangible values such as 'Australian', 'wholesome', 'family' and 'sporty', are all part of what we provide to our consumers. We must also provide value to our customers, the supermarkets that on-sell our products. They need a certain margin, and they need to have the product purchased by the consumer in sufficient numbers, or they will use that shelf space to sell another product.

In the USA over 80% of all new product introductions fail. Consider the cost of getting a new product to market and having it fail. Depending on the complexity of the product this can run into hundreds of thousands of dollars. This definitely does not add value to the business or to the consumer.

THE CHANGING MARKETPLACE

It is difficult and costly to get a new product onto the retail shelf. Retailers are expecting more in direct selling expenses, and shelf space is limited. The time a new product has to prove that it has a following is limited — sometimes to as little as six weeks. Product life cycles are also decreasing. Consumers are becoming tired of products more quickly, and so the time available to recoup the investment in new products is decreased. Consumers are more discerning and have a greater number of choices than ever before. Children, particularly, are constantly looking for the newest fad.

This makes profitable new product development seem extremely difficult.

Uncle Tobys, however, has a 90% percent success rate in new product introductions. What do we believe contributes to this success? There are many reasons, but the focus of this discussion is the use of sensory science. We see sensory science as a value-adding step in the process of developing and manufacturing food products, which in turn transfers that value to our customers and our consumers.

CONSUMER SCIENCES AT UNCLE TOBYS

Our Consumer Sciences group currently consists of four staff, five internal trained panels and one external trained panel, a small sensory testing centre, and a custom-designed computer package.

The internal panels are volunteers from our site. These range across all disciplines and all levels of management. The external panellists are volunteers from the local community. All panellists are trained for approximately three months before doing any 'real' work. Once their results are repeatable, and the panel results as a group meet our criteria, they begin work. The panels generally see a wide range of products, but we have smaller specialised panels for particular areas, for example, oats and rancidity.

We also use a children's panel which is semi-trained (ie, they are taught about descriptors and use of the scales). The children's panel is used for idea generation and idea screening. The Sensory Services group also conducts non-targeted consumer research. The group works very closely with our Marketing department, and with our consultants (as shown in Figure 6.1). This diagram shows how sensory science has an impact on almost every part of the business. If you took into account where our volunteers for the expert panels come from, we probably do interact with all parts of the business!

HOW DOES SENSORY SCIENCE HELP?

Within the business of new product development and manufacturing, sensory science assists in the decision-making process by removing subjectivity and providing fact. It is not a decision in itself, but provides information to support decision making. Any product developer will have memories of that special marketing person, General Manager or manufacturing manager who is absolutely certain that they represent the tastes of the target market, when their favourite food just happens to be fish milkshakes! Information from sensory science can enable you to escape those subjective circular discussions.

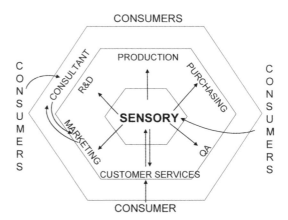

Figure 6.1
The interaction of sensory science with all aspects of the food manufacturing business.

CONSUMER NEEDS INTO NEW PRODUCTS

Uncle Tobys obtains information from consumers through three channels:

1   Customer Services   Complaints, questions, comments, suggestions and praise are channelled by consumers directly through our consumer information telephone line. This information is analysed and distributed.

2   Consumer Sciences   market research results, integrated trained panel data, concepts and focus groups, and consumer preference work are provided by this group.

3   Warehouse withdrawal data   This provides a picture of purchasing habits.

The process begins with Marketing, and Research and Development, agreeing on a Project Brief. An experimental plan is devised to cover a range of variables, for example, different textures, sweetness levels, and sizes.

Data from the trained panel is used to generate a sensory profile for each product. The sensory profiles are used to demonstrate that there is a sufficient spread of variables, and that there are statistically significant differences between products. Depending on the product, these samples may be run through a preference test, which allows the products to be refined before being sent to external consumer research. For example, if we have fourteen ideas for a new Rollups flavour, an incomplete block rank preference test would be conducted to reduce the products to the top four. This allows better targeting of a smaller number of samples. If the idea is very new, then the concept may be tested prior to the samples being trialed, enabling the assessment of the idea before the presentation of the product. This is important, because sometimes the concept is great but the product presented does not live up to the consumers' expectations.

Once the concept is refined, the key attributes driving liking are determined, and the trained panel data is integrated with the consumer information to obtain an optimum product profile. Research and Development can then begin to narrow the range of samples and processing methods, with the aim of closely matching the optimum.

Prior to Consumer Sciences being established, a product could return to external marketing research many times to get that optimum product. If funds were short, then the guy who likes fish milkshakes would make the final decision. The consumer can tell you what they want and what they like, but linking trained panel data with that information allows the product developer to understand what attributes are present, and at what levels. (Trained panels never do preference work.) Research and Development can make adjustments to formulations, and the trained panel can provide the information to indicate that the correct changes have been made. No further market research is required, which saves time and money in the development process. Consumer Sciences can turn around urgent information in four hours if necessary.

SENSORY SCIENCE IN MANUFACTURING

Trained and external panels are used to provide a wide range of information assisting other parts of the business, besides product development. The panels are used to profile raw materials for new and existing products to ensure the correct product profiles, and to provide a base-line profile in case raw materials need to change, either because of availability or cost. Profiles made of products in the pilot plant are compared to products produced in the factory to ensure that they are the same. Sensory data is then used in the set up of the specification limits around the critical consumer attributes, to help assess process capability. Panel data is also used in the shelf-life evaluation process, and in product evaluation sessions to determine attribute drift over time or processing problems. We are currently evaluating trained assessors using profiling for real time QC assessment. Consumer complaints about product performance are assessed by panels against the base-line profile. Finally, sensory data is used to optimise the product, once it is in production, for product performance and cost. Sensory Science is integrated into the Uncle Tobys business at many levels, and we are only just meeting the most obvious needs. There are many other areas requiring focus for the future.

MORE RESEARCH PLEASE

Some pure sensory research is also being conducted by our Sensory Services group. Understanding the mechanism of sensory perception is very important for the product developer. The Sensory Services group plays a role in educating and informing other parts of the business. However, I believe that insufficient attention is being paid to sensory science in Australia and little of the available funding is being channelled into this area. More pure research is required. The number one attribute driving liking is almost always taste, and we still understand very little about it.

CONCLUSION

Uncle Tobys uses sensory science as an instrument to provide data into its business systems — to develop new products fast, efficiently and to get it right the first time. That data is factual, scientific information that is used for effective decision making, ensuring that consumers get what they want in a cost-effective manner. There are problems and challenges to be overcome, even some disadvantages, but the benefits far outweigh these.

THE CHALLENGE

I ask you to take a close look at how you make your decisions regarding new product development and quality control, and see if the information you are basing those vital decisions on is sound and unbiased. Are you getting the results you expect? Are you getting those results consistently? Do you really understand your consumers' requirements? It is your consumers that you want to provide value to, and in doing so provide dollars to your company's bottom line. Take a close look at what sensory science can provide for you because it just may be that value-adding step that can provide the edge you are looking for.

UNDERSTANDING THE FOREIGN PALATE: SENSORY EVALUATION IN ASIAN MARKETS

G.A. BELL AND HAE-JIN SONG

INTRODUCTION

While food exporters stand to benefit from the continuing growth and diversification in Asian food markets, many still know very little about how to approach these markets. The determinants and nature of taste perceptions and preferences of consumers in overseas markets, particularly in the sizeable food markets of Asia, are poorly understood by food processors and product developers in countries outside Asia. Therefore, matching food ingredients to Asian tastes is deemed to be difficult by those wishing to engage in cross-cultural food export.

It is often believed that consumers in Asian countries are so difficult to understand that the task is not worth the effort. This leads to a 'dump or drop' strategy: if the product that is already formulated for the exporter's domestic market is not acceptable, then the project is dropped; if the product is acceptable it is simply 'dumped' into the market. This strategy does not succeed in the long term. This chapter aims to demonstrate that Asian consumers are 'fathomable', and that they can be studied with rigour, and with positive pay-offs.

It is surprising that very little cross-cultural research on the attributes of food acceptance is to be found in the scientific literature. Traditionally, food companies have used market research companies to tell them if a product is acceptable. Generally, they have not approached the problem from the angle of tailoring their product to the market, which calls for some systematic variation in the formulation of product prototypes. Previous information has been very specific to particular products. Very little is known of the information that is, presumably, retained exclusively by interested companies.

Cross-cultural comparison provides an understanding of one or more cultures in terms of the others. Australian sensory scientists have assisted many food companies to embrace the Asian opportunities from a knowledge base of the consumers in those markets, with particular emphasis having being placed on the perceptions, preferences, habits and culture of the consumers in those markets. A consistent and valid sensory technology has been applied across several markets in Asia, including Japan, Korea, Taiwan, Malaysia, Singapore and Indonesia. Asia is not a homogeneous market and hence it is necessary to assess each market individually.

SENSORY EVALUATION

Sensory evaluation attempts to discover which of the many sensations that occur when a piece of food is eaten, are the important ones in determining the overall liking of that food. When we eat something, we usually first see it and smell it, and then put it into our mouths. Once in the mouth we receive many other impressions: its flavour, its temperature, and its feeling in the mouth. These sensations can be measured by presenting the food in a controlled environment to people who are asked to rate each attribute of the food on a set of graphic scales. These rating scales provide measurements which, when averaged over thirty or more people, and treated with appropriate statistical tests, quantify the strengths and weaknesses of one's product and its competitors.

When dealing with food ingredients, systematic control of ingredients in prototype products or in model systems (such as solutions or gels), presented to panels of people under controlled conditions, can be used to determine whether one supplier's ingredient is stronger or better than another's. But what happens if people differ in their power to resolve differences? Common experience tells us that some individuals are clearly more sensitive than others. Some can describe what they are experiencing more accurately or meaningfully than others. These differences need to be taken into account if sensory evaluation is to be useful in choosing the best ingredients for products.

JAPANESE STUDIES

An extensive programme of cross-cultural sensory research was carried out from 1989 to 1994 on the Japanese consumer market by an international collaboration effort between Australian and Japanese scientists (see Prescott and Bell, 1995). The food habits of adults and children, and the complex distribution system for food in Japan were studied. Sensory evaluations of Australian products for companies wishing to enter the market were made, mostly benchmarking the sensory attributes of Australian products against foreign and local competition.

Notwithstanding the complexities of the Japanese distribution system, and of doing business in Japan, the consumer's role in a product's success is critical (see Bell *et al.*, 1992). Determinants of consumer purchase decisions are complex. Branding and price are important, as are perceived freshness and quality of most foods. Nevertheless, Japanese housewives rated *taste* as the most important determinant of their decisions to purchase processed food.

It was decided to challenge some of the existing beliefs about Japanese consumers by addressing generic questions, such as, are Japanese more sensitive than Australians to variations in taste? The first study examined sensitivity to variations in four basic tastes in model taste solutions: sweet, sour, salty and bitter. Panels of Japanese and Australian adults received solutions made up as either strong or weak concentrations. Each person was required to taste a series of pairs of the solutions. Each pair would be either of identical strength, or would vary slightly from the strong or weak strength standard. After tasting the two solutions, the subjects judged whether they were of the same or of different taste intensities. The object was to determine whether one group of people might be able to detect a smaller variation from the standard than the other group. The results (Laing *et al.*, 1993) showed that the Japanese were no more, or less, sensitive than the Australian group, for all four tastes, and for both strong and weak concentrations of the tastes.

As far as the basic tastes are concerned, we now know that the

Australian food exporter need not be over-concerned about the discriminatory power of the Japanese palate, since their 'taste machinery' is no more finely tuned than that of Australians. If the Japanese are highly discerning in their choice of foods, it must be for reasons other than their taste receptor sensitivity. How do their likes and dislikes for tastes compare with Australians?

In the second study, solutions of four basic tastes at six different concentrations were given to groups of adults in both countries. In addition, solutions of three so-called *umami*, or savoury tastes said to be well known in Asia, made up from the compounds MSG (monosodium glutamate), IMP (inosine 5'-monophosphate), and GMP (guanosine 5'-monophosphate), were included in the experiment. The subjects were asked to mark a scale representing the degree to which they liked or disliked each solution. For solutions of sucrose (sweet), sodium chloride (salty), caffeine (bitter), and IMP (one of the umami tastes) the subjects from each country showed identical relationships between the strength of the solution and the degree of liking or disliking. However, at strong concentrations of citric acid (sour) and two *umami* compounds, MSG and GMP, the two groups showed significant differences in the degree to which they disliked the solutions: Japanese disliked them less than Australians (Prescott *et al.*, 1992) (Figure 7.1). Both the similarities and the differences revealed by this study are interesting, but they are not in line with anecdotal evidence that the Japanese are necessarily quick to dislike strong tastes.

In the third study, Prescott *et al.* (1997) showed that manipulating taste components within a complete food shows up the differences in hedonic judgements familiar to many hard-worn food exporters. It compared sensory evaluations in both countries of a range of 36 sweet and 30 salty foods. Half the foods are commonly consumed in Japan, and half are widely consumed in Australian homes. Some foods were chosen because the type of food , such as strawberry jam, is common to both countries; but others were typical of each country, such as snack foods, or were modified for the Japanese market, such as 'Chokowa' breakfast cereal. In addition, the intensities of taste within the food context were manipulated. Australian food companies co-operated to provide their products made up to systematically varied taste strengths.

The results of this third study show that the intensity-sensitivity relationship from the first two studies is supported: that is, even in the complex food context, sweetness and saltiness intensity are judged equivalently by subjects from the two countries, but, in the context of

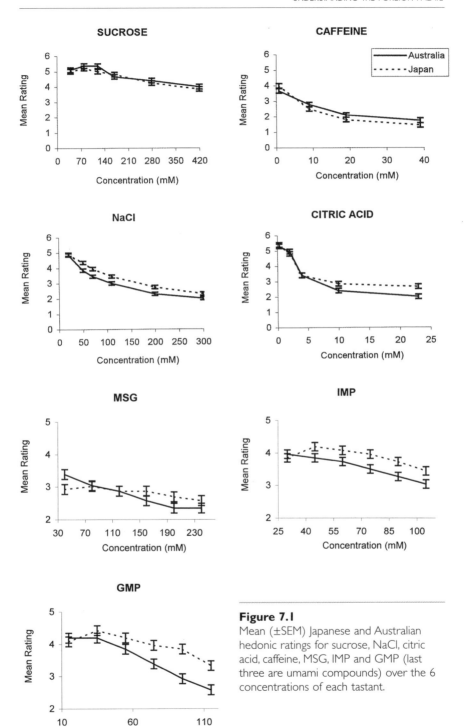

Figure 7.1

Mean (±SEM) Japanese and Australian hedonic ratings for sucrose, NaCl, citric acid, caffeine, MSG, IMP and GMP (last three are umami compounds) over the 6 concentrations of each tastant.

food, the liking of tastes by the two cultural groups showed significant differences. Hence, these studies show that hedonics of certain basic tastes *in foods*, but not in model solutions, differ between the Australian and Japanese cultures. This highlights the importance of food context in taste judgement. Food context is especially crucial in cross-cultural taste research, where it is possible that the same foods are consumed in different contexts by different cultures.

Over the past ten years, sensory analysis has been applied to a large number of Australian food products, providing data for their reformulation, and contributing to the successful marketing of products modified for Japan. The competitive position of food exports to any foreign country can be improved by scientific study of consumers' sensory perceptions in export markets.

It might be the case that human populations have differing proportions of people who are more sensitive to some tastes, fragrances and flavours (including pungency). This could account for the degrees to which the Japanese and the Australians differed in their liking of the solutions in the Prescott *et al.* (1992) study: a small number of strong dislikers could bias their group's scores. Research is gaining momentum (inspired greatly by the research of Linda Bartoshuk at Yale) to see if people within a culture, as well as those from different cultures, may not have sub-groups who are particularly sensitive to some chemosensory stimuli. The ideas and some of the recent research on the role of 'supertasters' in ethnic tastes are reviewed in a useful article by O'Donnell (1997).

In Japan and elsewhere in Asia, there is growing concern about health issues and food, particularly among the affluent young, and about the special dietary needs of specific market segments — particularly by mothers with young children and the increasing numbers of aged Japanese. These trends create pressure on product developers to come up with healthy, so-called 'functional' foods, that are acceptable to new categories of consumers. The ingredients for functional or health foods are often not very palatable, and therefore the healthy, and sometimes bitter, ingredients need to be balanced by others which make the product more palatable. It is rarely the case that demand for an unpleasant tasting food will be sustained only because of the purported health benefit.

INDONESIAN STUDIES

Indonesia is an 'emerging market', and one of several highly populated Asian countries which, notwithstanding recent economic problems, have had steady, strong economic growth for most of the past decade.

As a result, people with buying power have grown in numbers. At the highest end there are market strata that have never existed before to any significant extent. In the past ten years, between two and five million Indonesians have gained incomes comparable to those of the affluent middle-classes of Europe and North America. They are the so-called 'A+' consumers. With steady economic growth there could be 50 million A+ consumers by the end of the first quarter of the next century. There will be corresponding increases in numbers of consumers in other market segments (A, B, C, and D), thus presenting significant markets for consumer goods, including processed food and beverages. Two studies are reported here which have looked at Indonesian consumers in the highest segment (A+), and in segments down to 'D' level. These studies have identified drivers of purchase for different food categories, including cultural, gender and age factors.

A study of the A+ consumers, conducted in Jakarta in 1996, consisted of three structured discussion sessions (focus groups) of ten to twelve participants, and a questionnaire given to 114 adults who had not been in the focus groups. The results are reported in detail by Easton *et al.* (1997), and in this volume again by Easton.

Another recent Indonesian study, conducted in Bandung (Song, 1997), consisted of four focus groups, a questionnaire on food usage and sensory evaluations by panels of 30 women on a number of 'off the shelf' products from Australia. In addition, several prototype products, including some in traditional style, were evaluated. Consumers from two age groups in the A–B and the C–D market segments took part in the study: Information was obtained from 'young adults' (17- to 24-year-old females) and 'mature adults' (25- to 40-year-old females).

Among the factors assessed was the priority people put on attributes of a range of foods when making purchasing decisions. Taste was again one of the main drivers, in both age groups and both market segments. Price was more important to younger and less affluent consumers. Brand was more important to older and more affluent consumers. Imports from various countries are considered of high or low quality depending on country of origin. Australia, USA, Japan and Europe scored equally highly. India and the Philippines scored least well.

Indonesians placed a high importance on spice in their diet, particularly chilli. Of the savoury products evaluated by the sensory panels, most were considered not spicy enough. However, some 'hot' products, formulated as 'medium-hot' for North American consumers, were considered 'too hot'. This finding was surprising in light of the high levels of 'burn' tolerated in many traditional Indonesian dishes.

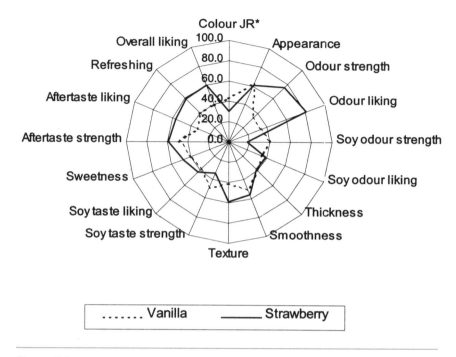

Figure 7.2
Comparison of attribute ratings for strawberry and vanilla flavoured beverages.

Of the flavours used in the various products, there were some that were clearly not appropriate. Two examples were the choice of herb flavours in bread and the use of vanilla in beverages (Figure 7.2). Both reminded consumers of medicines, and were clearly rejected in the food evaluations. The less familiar the food was the more likely it was to be rated poorly on overall acceptance.

The Indonesian studies emphasise the need for adequate knowledge of the target market to be obtained and for products to be tailored for it, particularly heeding the importance of taste in purchase decisions. Sensory evaluations of potential products for the emerging Indonesian markets are highly recommended, prior to full market trials or product launches.

THE EXPATRIATE'S PALATE

Within a culture, however, tastes can change with the affluence of the consumer, and with age and social reference group. This was demonstrated in a cross-section of Japanese consumers in 1988, in which the favourite foods of the young consumers reflected the newly introduced products such as fruit juice and ice cream, while older consumers stuck to traditional foods, with some exceptions (such as French fries) (see Bell *et al.*, 1992).

Doing research in Asia, particularly Japan, requires resources that add cost to a project. The presence of Asian communities in many western countries is often regarded as an opportunity to prepare products for an Asian market without having to test the products in that market. However, doing so runs the risk of tailoring a product for a market which does not correspond with the true target market: among the reasons for this is that the expatriate community is more affluent and selected than the main population and because palates do change after time out of the 'mainstream'.

In addition, the perceived expectations of the experiment, and the preconceptions of the quality of a product very easily bias people's taste responses. If a subject in an evaluation panel thinks he is tasting foreign food, or his national product is in the comparison, the results will come out biased.

Tastes change when a community establishes itself outside the country of origin. This was demonstrated in an unpublished study of noodles in Korea and in the Korean community in Sydney. The two groups showed poor concordance on ranking the quality of 12 types of noodles. Accordingly, food companies are advised to test products in the country of interest, and to employ trained staff from that country as the interface between the consumers and foreign sensory experimenters.

To make sensory evaluation methods pay off in Asia, apart from adequate training of personnel, the methods need to pay due respect to the translation of the instructions, questionnaires and psychometric scales into the language of the sensory panellists. All the staff in the evaluation session should be of similar ethnicity to the panel and the instructions need to be communicated in a friendly manner, which inspires genuine interest and co-operation from the subjects.

The result of meeting Asian consumer demands is often surprising: there might be no modification needed to one's product, or, on the contrary, it might need major redesign, or indeed it might be best not to attempt to export any form of a particular product.

How a food export might perform will depend on how well it meets the sensory expectations of the consumer. Naturally, this will form part of the complexity of purchasing behaviour, along with price, packaging and product image (see Prescott and Bell, 1995).

There are many things to consider, including a general understanding of how and where the export product might be used, how it is distributed, and so on (see Bell *et al.*, 1992; Prescott and Bell, 1995; Easton *et al.*, 1997). However, if the taste of the product is not right, the buyer will not return after the first trial. The first purchase is often made on the basis of curiosity, and gives a false impression of the market's response to it.

Exporters need, most of all, to suspend personal opinions (and those of the managing director's), beware of experts and trust the messages that good quality consumer behavioural research can deliver. It makes good sense to use information on the target consumers' taste perceptions, attitudes and preferences to guide the development of products for export.

REFERENCES

Bell, G.A., Ng, F., Waring, J. and Vereker, M. (1992) *Exporting to Japan: A Guide to the Food Market and Distribution System.* CSIRO, Sydney.

Easton, K., Bell, G.A. and Ng, F. (1997) *Exporting Food to Indonesia. A Guide for Australian Small to Medium Enterprises.* CSIRO, Sydney.

Laing, D.G., Prescott, J., Bell, G.A., Gillmore, R., James, C., Best, D.J., Allen, S., Yoshida, M., and Yamazaki, K. (1993) A cross-cultural study of taste discrimination with Australians and Japanese. *Chem. Senses*, 18(2), 161–168.

O'Donnell, C.D. (1997) Formulating products for ethnic tastes. *Prepared Foods*, Feb. 1997, 36–44.

Prescott, J., Laing, D., Bell, G.A., Yoshida, M., Gillmore, R., Allen, S., Yamazaki, K., and Ishii, R. (1992) Hedonic responses to taste solutions: a cross-cultural study of Japanese and Australians. *Chem. Senses*, 17(6), 801–809.

Prescott, J. and Bell, G.A. (1995) Cross-cultural determinants of food acceptability: Recent research on sensory perceptions and preferences. *Trends Food Sci. Technol.*, 6, 201–205.

Prescott, J., Bell, G.A., Yoshida, M., O'Sullivan, M., Korac, S., Allen, S. and Yamazaki, K. (1997) Cross-cultural comparisons of Japanese and Australian responses to manipulations of sweetness in foods. *Food Qual. Pref.*, 8(1), 45–55.

Song, H-J. (1997) *Gustatory and olfactory perceptions and preferences of Indonesian consumers.* Thesis, BSc. Food Science and Technology, The University of New South Wales, Sydney.

A SMELL TEST BASED ON ODOUR RECOGNITION BY JAPANESE PEOPLE, AND ITS APPLICATION

S. Saito, S. Ayabe-Kanamura, T. Kobayakawa,
Y. Kuchinomachi and Y. Takashima

INTRODUCTION

Saito and Arakawa (1995) showed that recognition of odours may be affected by the experience of odour during development. If the odour recognition of Japanese people is influenced by Japanese food and lifestyle, then the Japanese cognitive space of odours should be different from that of American and European people. When Japanese people try to express or identify one odour using linguistic descriptors, the descriptors will be extracted from the Japanese cognitive space of odours. Thus, it is important to think over a unique cognitive odour space for the Japanese in developing a test of odour identification for the Japanese.

The University of Pennsylvania Smell Identification Test (UPSIT) (Doty and Shaman, 1984), is a very useful test to measure the identification performance of odours. Some of these odours, however, may not be suitable when the test is applied to Japanese people. The test contains several odours that Japanese people have not experienced and cannot even name, and conversely it does not contain several odours that are familiar to Japanese.

In this study we first classified the odours in Japanese cognitive space and chose the sorts of odours for a Japanese smell test from the results of the classification, adding the results of other studies. Then we searched out an appropriate odour material for each odour, and made a smell test by microencapsulating the odour material and printing it on a small piece of paper. This test, named STAUTT3 (Smell Test of AIST, University of Tsukuba and Takasago Int. Corp.), was investigated to confirm its reliability and applied to older people to study their olfactory characteristics. Subjects in this study were informed about the nature of the experiment, and agreed to become subjects. The plan of the present study was acknowledged and authorised by the National Institute of Bioscience and Human Technology, Japan.

CLASSIFICATION OF ODOURS IN JAPANESE COGNITIVE SPACE

To classify the odours of Japanese cognitive space, 98 descriptors of odours were used. The 98 descriptors were selected from a larger group of 204 (Saito et al., 1997) by the following three criteria: (1) odour descriptors commonly known to Japanese people, (2) descriptors of specific odours, (3) similar descriptors that could be grouped into one category. Some of the odours unique to Japanese recognition were warm vinegar rice (warm sushi rice), Japanese horseradish (wasabi), miso (soybean paste), soy sauce, roast dry cuttlefish, grilled eel with soy sauce, dry seaweed, seaweed boiled in soy sauce, tangle flakes, dried sardines, fish stock, ginkgo nut peel, salted rice-bran paste for pickling (nukamiso), green tea, and tatami.

A pile of 98 small cards (6 cm x 9 cm), on which one descriptor of an odour was written, was presented to each subject, who was asked to make groups of cards by the similarity of odours shown by the written descriptor on each card. Subjects classified the cards twice by the following two different criteria: (1) The number of groups is arbitrary, (2) The number of groups is as few as possible. We added the second criterion because most subjects had a tendency to classify into too many groups using only the first criterion.

The subjects were 26 healthy Japanese volunteers (16 women and 10 men) who lived in Japan. They were students, housewives, or institute personnel with no history of olfactory impairment. Their ages were from 18 to 49, with 12 subjects under 30 years old and 14 subjects over 30. Fifty sets of data classifying odour groups (the data from one subject were rejected because of an incomplete answer) were subjected to a cluster analysis (Johnson, 1967).

The classification of odours in Japanese cognitive space is shown schematically in Figure 8.1, divided into four cluster levels of same similarity by four concentric half ellipses. The 98 odour descriptors were clustered into 26 groups in the largest concentric half ellipse in Figure 8.1, then into 16 groups in the second largest one, then into eight groups in the third largest one, and then four groups in the smallest one. Compared with the six odours (flowery, fruity, putrid, spicy, burned, resinous) represented in Henning's odour prism (Engen, 1982), three kinds of odours (spice, rotten, and plant) were represented at the cluster level of eight groups in this study. Burned odour was represented at the cluster level of 16 groups, and flower and fruit at the cluster level of 26 groups. Japanese people clustered these odours into one group as sweet odours. This means that flower and fruit odours are not clearly distinguished by Japanese people compared with European people. The reason might be that most Japanese flowers and fruits do not have a strong fragrance. Conversely, groups represented by flavor, fishy, and chemical odour, not mentioned in Henning's odour prism, were big clusters in this study. These clusters reflect the unique Japanese food culture.

DEVELOPMENT OF A SMELL TEST BASED ON ODOUR RECOGNITION BY JAPANESE PEOPLE

To develop a test based on odour recognition by Japanese people, odour qualities were selected from the results of the cluster analysis described above, the odours used in the T&T olfactometer (Takagi, 1989), and the odours recalled by older people (Tanigawara *et al.*, 1994). Thirty-five odours were chosen for consideration (Table 8.1). Odour substances matched to each selected odour quality and their appropriate concentrations were decided, and microencapsulated and printed within a circle of 25 mm diameter on a small piece of paper (6.5 mm x 3.5 mm).

For some odours (seaweed, faeces, mould, gasoline, joss stick, and boiled rice), it was difficult to make artificial odours, and for some (vinegar, sake, sweat, cigarette, and green), it was difficult to microencapsulate the odour materials. Smell test cards of 22 odours were made (the first 22 odours in Table 8.1). Subjects scratched the circle of each test card 10 times by the head of a pen, and sniffed the test odour. The identification of odour quality was tested by having the subjects choose the correct odour from four alternative odour

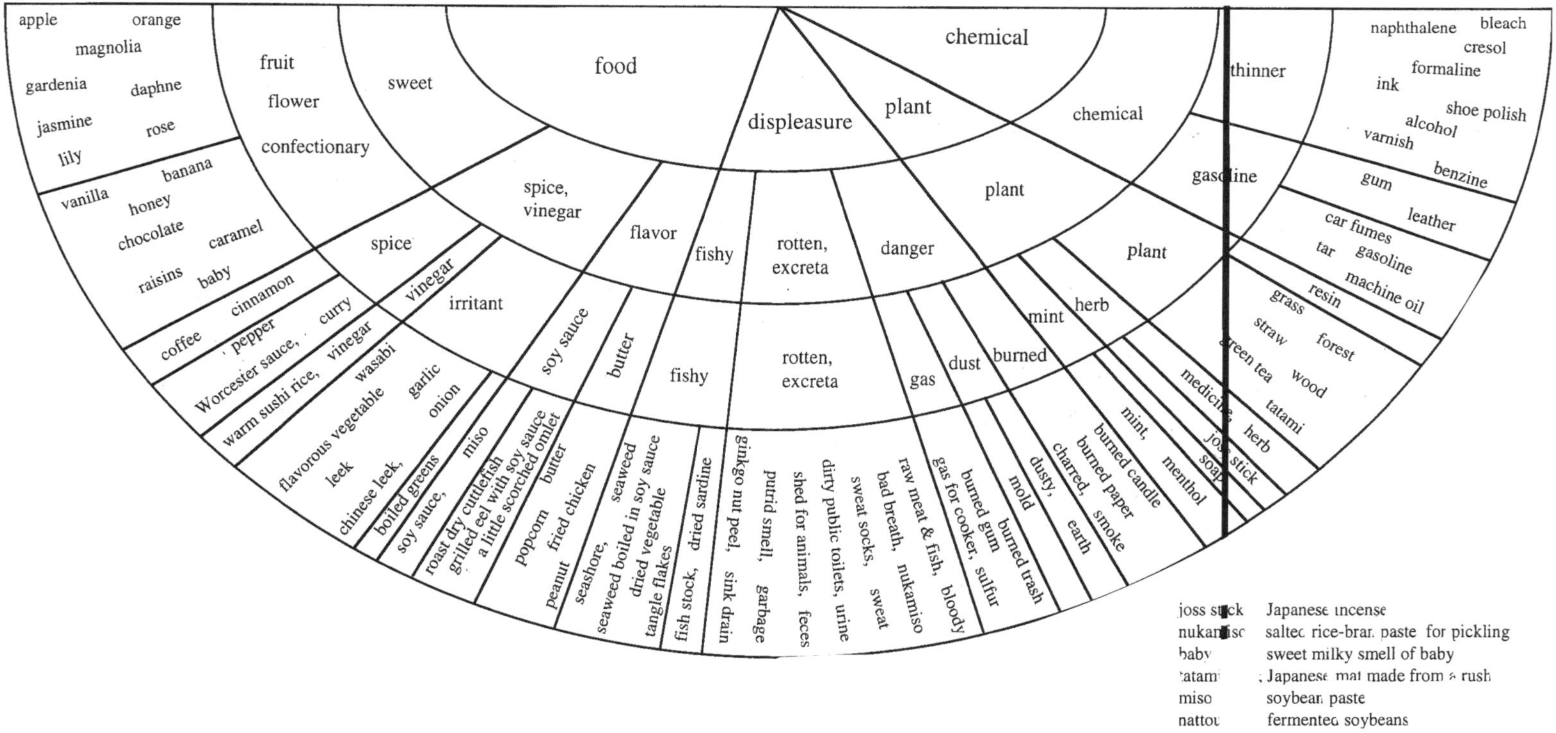

Figure 8.1
The classification of odours in Japanese cognitive space shown schematically. Four cluster levels of same similarity obtained from a cluster analysis are shown by four concentric half ellipses. Ninety-eight odour descriptors were classified by four cluster levels of same similarity, and each cluster was named by typical or representative odour. (Saito et al., 1994)

joss stick   Japanese incense
nukamiso   salted rice-bran paste for pickling
baby   sweet milky smell of baby
tatami   Japanese mat made from rush
miso   soybean paste
nattou   fermented soybeans

descriptors. Alternative descriptors for each odour were prepared in several different levels of dissimilarity between odours, depending on the distance between clusters. It was also possible to measure the perceived intensity using a six-point scale from no odour to very strong.

The reliability of STAUTT3 was examined on the following points:

1   Consistency of intensity in every card The intensity of 64 different cards was evaluated from eight to ten times by five subjects. The differences in intensity from the mode intensity for each subject was from -1 to 0.5 in all cards, and from -0.5 to 0.5 in 62 cards (97% of all cards).

2   Stability of odour release for different numbers of scratches The gas volume of an orange odour card was measured successively after scratching 10 times or 40 times, using a semiconductor gas sensor. No remarkable difference was observed between the two conditions (Figure 8.2 on page 81).

3   Appropriateness of odour substance, its concentration and descriptors Eighteen subjects identified the quality of odour, and evaluated the intensity of 22 odour cards to check the appropriateness of the odour substance, its concentration, and its description (given the four alternative descriptors offered in the identification test). Seven in 22 odours were identified at 100%, 12 odours above 90%, 16 odours above 70%, 21 odours above 60% and all above 56%. These results indicate that most odour cards were well developed, considering the chance level of 25% for four choices.

APPLICATION OF STAUTT3 TO OLDER PEOPLE

To investigate the characteristics of the olfaction of older Japanese people, STAUTT3 was applied. Prior to this test, 318 older people were asked if they had experienced any decline in olfaction, as part of a questionnaire on life quality. Forty-eight (15.1%) of the group of 318 reported some decline in olfaction.

STAUTT3 was applied to 21 older people (9 women, 12 men), whose average age was 73 years. One of them reported a decline of olfaction on a questionnaire. The same test was applied to 21 middle-aged women (average age, 35 years). The odour card was scratched by an experimenter, in the case of the older group. Inter-stimulus interval was four minutes. Four odours (camphor, caramel, cresol, and bad breath) were excluded from analysis because of deterioration of the odour. The data of 18 odours were analysed in terms of the percentage of correctly identified odours, and the perceived intensity of the odours.

The average number of odours correctly identified by the older group was 10.5 (58.3%) (SD = 3.8, 21.1%). This was significantly smaller than the 14.4 (81.7%) (SD = 1.7, 9.4%) correctly identified by the middle-aged group (t(40) = 4.52, p < 0.001). The average

Table 8.1
Odours selected for a smell test of Japanese people

Odour quality	Criterion	Odour quality	Criterion	Odour quality	Criterion
musk	T	curry	C	seaweed	C
canned peach	T	butter	C	mould	C
nattou/sweat socks	T, C	gas for cooker	C	gasoline	C
camphor	T, C	putrid smell	C	joss stick	C
bad breath	T, C	menthol	C	boiled rice	Q
garlic	T, C	smoke	C	vinegar	T, C
caramel	T, C	Japanese orange	C, Q	sake	Q
cresol	T, C	wood (Japanese cedar)	C	sweat	Q
rose	T, C	hinoki (Japanese cypress)	Q	cigarette	Q
banana	C	Japanese traditional ink	Q	green	Q
soy sauce	C	milk	Q	faeces	Q
apple	C	perfume	Q		

criterion T: selection from odours used in T&T olfactometer
criterion C: selection from the result of cluster analysis
criterion Q: selection from the odours recalled by older people

perceived intensity in the older group was 2.75 (SD = 0.60), significantly lower than the 3.14 (SD = 0.34) reported by the middle-aged group (t(40) = 2.53, p < 0.05). The number of correctly identified odours in the older group ranged from 2 to 17, compared to 11 to 17 in the middle-aged group. Hence, individual differences in the older group were larger than in the middle-aged group. The subject who reported some decline of olfaction showed lower performance (39% in percentage of identification and 1.0 in averaged perceived intensity). However, most subjects of lower performance did not report any decline in olfaction.

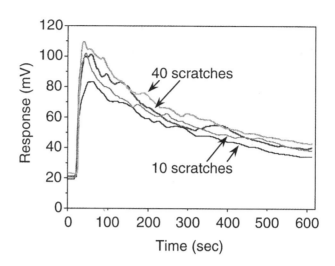

Figure 8.2 Stability of odour gas volume by different number of scratches (in the case of a orange odour card). The gas volume (vertical axis) was measured (using a semiconductor gas sensor) after scratching 10 times or 40 times successively. (Saito *et al.*, 1994)

CONCLUSION

To develop a smell test suitable for Japanese people, odour qualities were selected from the various odours recognised by the Japanese, using various criteria, including cluster analysis. An appropriate odour material was found for each odour quality, then microencapsulated and printed on a small piece of paper. Identification performance of odours and perceived intensity were measured by this test. The reliability of this test was examined and the validity was practically confirmed by an application to older people.

ACKNOWLEDGMENTS

We thank Dr. Miyano and Mr. Utsugi for their help in the application of cluster analysis.

Figures 8.1 and 8.2 were reproduced with permission from Saito *et al.*, 1994.

REFERENCES

Doty, R.L. and Shaman, P.M. (1984) Development of the University of Pennsylvania smell identification test: A standardized microencapsulated test of olfactory function. *Physiol. Behav.*, 32, 489–502.

Engen, T. (1982) *The perception of odours*. Academic Press, NY, pp 8.

Johnson, S.C. (1967) Hierarchical clustering schemes. *Psychometrika*, 32, 241–254.

Saito, S. and Arakawa, C. (1995) Does pleasantness of a certain odour change by one's growth environment? *Chem. Senses*, 20, 102–103.

Saito, S., Ayabe-Kanamura, S. and Takashima, Y. (1994) A smell test based on odour classification of Japanese people. *Jpn. J. Taste Smell Res.*, 1–3, 460–463 (s327–s330).

Saito, S., Iida, T., Sakaguchi, H. and Kodama, H. (1997) Description of quality of pollution odours(Japanese). *J. Odour Res. Engineer.*, 28, 32–43.

Takagi, S.F. (1989) Standardized olfactometries in Japan — A review over ten years. *Chem. Senses*, 14, 25–46.

Tanigawara, C., Watanabe, K., Satoh, S., Saito, S., Ayabe, S., Matsuzaki, I. and Oda, S. (1994) Odour arousing affections of the nostalgia to the aged Japanese. (Japanese). *Jpn. J. Ergonom.*, 30, 51–56.

INDONESIA:
TASTE PREFERENCES
OF A+ CONSUMERS

K. EASTON AND G.A. BELL

INTRODUCTION

International trade has become a key focus for food companies wishing to expand their markets. At this point in time, many of the ASEAN countries (eg, Malaysia, Indonesia, Singapore, Thailand, Brunei and the Philippines) are experiencing a period of economic upheaval, which may affect food exporters' short-term plans for market expansion. However, the long-term view is towards economic recovery, and this still requires analysis of gaps in each country's domestic food supply, as well as a keen grasp of the niche markets available to imported food products.

Many international companies develop con- .sumer products tailored to the Asian markets in an effort to cater to the hedonic preferences of that consumer group, and in doing so face a variety of problems. These include such issues as the need to identify which cultural (Rozin and Vollmecke, 1986) and religious traditions may influence food choice, and the suitability of a type of food to its destination country. The observation of Islamic dietary laws in Muslim countries is an obvious example of how religious traditions guide food choice, while appropriateness of the type of food to a country is confounded by a variety of factors, including the identification of preferred flavours in various foods (Prescott et al., 1998), and familiarity with the food product and its uses (Pliner and Pelchat, 1991).

Companies entering the Asian market are advised to selectively target particular groups of consumers or regions in the chosen country (Tyler, 1998). Indonesia is a country of almost 200 million people, where the top 1% of the population, in terms of affluence, are termed the A+ consumers. They hold much of the purchasing power, and have previously been exposed to Western-style foods through travel and an increasingly cosmopolitan lifestyle. Their food consumption patterns are changing, along with their shopping habits, as they adopt the convenience and comfort of modern supermarkets.

When considering any new market it is necessary to gather as much information as possible on the habits and preferences of its consumers. Despite the readily available evidence of changes to shopping facilities in larger cities in Indonesia, and the effect of these changes on food purchasing habits (Datamonitor, 1997), success in entering this market also depends on having a thorough understanding of the cultural and religious sensitivities of the chosen consumer group. These have strong influence on food choices and taste preferences. Previous cross-cultural studies have demonstrated diversity in sweet or salty preferences between different cultural groups (Lundgren et al., 1976; Lundgren et al., 1978; Druz and Baldwin, 1982; Bertino et al., 1983; Aminah 1995; Prescott et al., 1997), and that preferences for levels of sweetness or salt are influenced by familiarity with the foods containing them (Bertino et al., 1982; Prescott et al., 1993; Laing et al., 1994). Other studies have examined the influence of prior exposure to particular foods on subsequent food and taste preferences (Moskowitz et al., 1975; Bertino and Beauchamp, 1986; Prescott and Khu, 1995). However, the majority of studies have concentrated on tastants in solution (Moskowitz, 1971; Johansson et al., 1973; Moskowitz et al., 1974, 1975; Lundgren et al., 1976; Yamaguchi et al., 1988) rather than on consumer structures such as socio-economic level, attitudes, taste preferences, eating habits and purchasing criteria (Tyler, 1998) that determine the demand for branded, processed foods. It is these consumer factors which ultimately affect the food industry's evaluation of the branded segment of the processed food market in terms of price, quality, convenience and added value.

A study conducted in Jakarta in March 1996 (Easton *et al.*, 1997) with groups of A+ Indonesian consumers was aimed at developing a better understanding of the consumers' perspective. The first part of the study consisted of three structured discussion sessions (focus groups) of 10 to 12 participants, and the second component was a questionnaire administered to 114 adults who had not been in the focus groups. The results of this study provided insight into understanding the flavour palate and quality expectations of this A+ consumer group, as well as information on their shopping habits, favourite foods, use of convenience foods, and their attitudes to the attributes of foods they consider most important.

THE FOCUS GROUPS

The focus groups were conducted as three separate structured discussion sessions, each led by the same moderator. The participants were all A+ consumers, and were divided into groups according to their age: 'young adults' (17- to 24-year old males and females), 'middle adults' (25- to 34-year old females), and 'older adults' (35- to 54-year old males and females). Each session was conducted in Bahasa Indonesia (the official Indonesian language), and produced qualitative information about the following points of interest.

Shopping habits

The young adult A+ consumers were typically senior students whose parents were responsible for family shopping. The two older consumer groups were the primary decision makers for weekly grocery shopping, usually done in supermarkets. Fresh seafood and vegetables were regularly purchased in the traditional local markets (not the Wet market), on a daily basis when possible, as freshness was important to these consumers. The A+ consumer performed this task if time permitted, but if they were working, it was done by the household maid.

Food preparation in the home

Basic food preparation at home was largely Asian in style, prepared by the family maid. The middle and older groups preferred traditional-style foods for themselves, but would prepare Western foods when their children requested it, or would take them to Western-style restaurants and fast food franchises. Restaurants were the main venues for social eating. The younger consumers displayed a shift to Western-style tastes, with part of the appeal due to a liking for new tastes, and part associated with the attraction of a fashionable location.

However, the new tastes had not completely displaced the old, as they still bought traditional foods at school canteens and street stalls.

Traditional snack foods

Although the traditional *kaki lima* (street food vendors with barrows) cater to the masses, and are assured of a continuing role in the Indonesian food scene, the A+ consumers placed a high value on the sensory qualities of taste and smell associated with the traditional foods they supplied. *Warungs*, or stall restaurants, are another convenient source of traditional foods, but as with the *kaki limas*, the consumers choose the vendors carefully, as hygiene is important (Winarno, 1992). Often, they would not purchase from the *kaki lima* when their children were with them, because of concerns about putting the children at risk from food poisoning. These sources of 'street food' are estimated to provide around 30% of the daily food intake in all urban households in Indonesia (Bijlmer, 1992; Hubeis and Marlin, 1992).

Social eating

Social eating among the young adult consumers was a mixture of Western-style snacks and fast foods, balanced against a continuing allegiance to traditional snack foods, which were still perceived as being both tasty and conducive to a relaxed atmosphere. Italian foods and pizza were very popular for consumption away from home, but basic eating was still Asian in style, and the menu at home was fairly traditional because it was generally prepared by the family's servant.

The younger consumers were a little more relaxed about snacking in public than the older groups, some of whom still felt reluctant to eat while strolling. Nevertheless, the young ones did discriminate between the social acceptability of small food items versus larger ones (eg, doughnuts were too big). However, the food service 'drive through' removes these constraints, as eating in the car is quite acceptable.

Among the older groups, dining together as a family happened about 50% of the time due to work commitments, but special events such as birthdays were considered important, and most people make an effort to celebrate the occasion together.

Attitudes to religious strictures concerning food (the Halal factor)

About 87% of the Indonesian population professes Islam as their religion, but there is a comparatively much higher percentage of non-Muslims (around 50%) in the A+ class. The Islamic dietary laws include abstinence from pork products, guidelines for meat slaughter practices, and other rules relating to consumption of foods and drinks defined as meeting the requirements of their religious principles (Slamet, 1993). Our study confirmed that Halal certification on meats, on processed meats like sausage, and on such items as biscuits and canned foods is important to the Muslim consumers surveyed — if there's no Halal label, they will not buy the product. Foods such as *sambal* (chilli sauce), cakes, sauces, jam and sugar are not the kinds of foods which these consumers would check for Halal labeling as they do not usually contain any products considered harmful (*haram*). Non-Muslim consumers were not concerned with Halal acceptability for foods, but were interested in ingredients and expiry dates on labels.

Attitudes to branding and packaging

Branding was important to the mature A+ consumers in this study, for identifying which products would be purchased again and which to avoid. Brand loyalty was strong in everyday items like cooking oils, *sambal*, oyster sauce, and cooking spices, but less adherence was felt for products like chocolate and candy, where expiry dates were checked and there was interest in sampling new brands. The two older consumer groups felt that quality packaging on confectionery was both tempting and indicative of the quality of the contents, but the young group felt the pressures of a pocket-money income. They liked brightly coloured, eye-catching packaging, but wanted value for money in comparable products. Easily damaged packaging was disliked, as were oversize packs for snack foods. Most consumers in the two older groups paid close attention to their children's preferred brands.

Prepared foods — canned, fresh and frozen

Prepared spices, condiments and instant noodles were frequently purchased, but seldom frozen foods (apart from ice cream, cakes, chicken nuggets and sausage). These A+ consumers expressed concerns about the quality of taste and the presence of preservatives in canned and frozen prepared foods. When they did purchase these goods, it was usually for novelty-value so they could try new products, for convenience when supermarkets were closed, or as stand-by pantry supplies. The presence of a servant for the family reduced the necessity for prepared foods.

Traditional foods

Dishes such as *gado-gado* (vegetable salad with peanut sauce), *satay* (marinated grilled meats), *sayur* (vegetables cooked with coconut milk), *sop* (clear soup) and snack foods like cakes and finger-foods were regarded as traditional by the A+ consumers. There were preferred vendors and businesses specialising in these prepared mainstays of Indonesian cuisine — preferred because the flavour of the food was good, and their level of cleanliness reduced concerns about the short duration of shelf-life for these products. Traditional snack foods are never frozen, as the taste diminishes in the process, but they are a regular part of life, although not necessarily purchased every day. Boredom with flavours is never an issue, as there is such a wide variety of snacks available that there is little chance of repetition.

Healthy food

Health- and nutrition-consciousness was taking a high profile with these A+ consumers. Fish was perceived by the middle adults as being healthier than meat, while all groups said milk/yoghurt, vegetables and fruit were all good for them. However, the younger consumers placed tasty and satisfying foods higher in importance than healthy ones. Children's preferences dictated the choice of fruit purchased within the family, but country-of-origin of the fruit was not important to the consumers.

Price of food

The price-consciousness, and susceptibility to economic swings, of lower income Indonesian consumers is not reflected in the purchasing habits of the upper income bracket, who along with most expatriates will pay two to three times the price of goods in the United States (USDA: FAS Online). Our study confirmed that, as a whole, the focus group participants were not particularly price-conscious, although the young group was price-conscious when comparing similar product categories (eg confectionery) because they relied on pocket-money.

Imported foods

Quality and taste were the main criteria for choosing imported foods, rather than the country-of-origin. Certain goods, like chocolate or cheese, were identified for quality by their source, but imported foods in general filled a niche rather than a primary place in food purchasing habits.

THE QUESTIONNAIRE

The questionnaire (Easton *et al.*, 1997) provided numerical data on food spending, shopping habits and favourite foods, drinks and cuisine preferences, as well as measuring the priorities people put on attributes of a range of foods when making purchasing decisions. Taste was one of the main drivers, as well as quality and freshness, which also rated highly. Nutrition was very important to these A+

consumers, but not as important as the sensory aspects. Halal acceptability was extremely important to the Islamic A+ consumers, and had the highest importance of all attributes when choosing meat. Product labeling in all categories was important to Muslim consumers, reflecting the necessity to demonstrate Halal integrity for the Islamic majority of Indonesia.

In our study the household expenditure on food was broken down — into 60% spent on food consumed at home, and 40% on food consumed away from home. The A+ consumers were the main purchasers of the household's grocery items, with only a very small percentage (5%) leaving the responsibility for purchase decision-making to the servants. The Islamic A+ consumers tended to eat out a little more, had a slightly higher income level overall, and shopped more frequently per week than other groups.

Favourite foods at breakfast were recorded as *roti* (bread) with a filling or spread, *nasi goreng* (fried rice with egg), or noodles. Lunch and dinner featured rice strongly on the menu, as many Indonesians feel that meals without rice are not 'real' meals, more like a snack. Steamed rice was a favourite at lunchtime, as well as for the evening meal, but *nasi goreng* was almost equally favoured at night. Many of the same style of dishes were used for both meals. Soups, noodles, fried chicken, fragrant savoury dishes such as *satay* and *rendang* (a dry curry), and vegetables steamed or in a sauce (stir-fried, or fresh as in *gado-gado*), or soup, were some of the most popular foods in this group of consumers.

Chocolate topped the list of favourite sweet foods (as it does in many other countries). Traditional sticky cakes (in many varieties, called *kue*) were also highly popular, as were sweet stewed fruits, often cooked in coconut milk. Oranges, *sayur asam* (vegetables in sour tamarind-flavoured soup) and *asinan-manisan mangga* (literally, sweet-and-salty mango) were the most liked sour foods, along with *empek* (fishcakes with sweet and sour sauce) and sour confectionery. Salted fish, *asinan* (crisp fruit and vegetables in sweet and sour dressing) and cheese were the favoured salty foods. *Sambal*, *rendang* and *rujak* (spicy fruit salad) lead the responses for the most preferred spicy foods.

Coca-Cola was named the favourite cold drink, but local iced fruit drinks were still very widely used for refreshment, and in total responses outnumber the proprietary soft drinks. *Gado-gado*, *rendang* and various rice dishes proved to be the most popular traditional Indonesian meals, but the cuisine is so varied it is difficult to assign dominance to any of the other responses for favourite traditional meals.

Hamburgers, steak, fried chicken and pizza have acculturated well, each finding a strong niche in the 'most favoured' ranks of non-traditional meals.

Indonesians place a high importance on spice in their diet, particularly chilli, which is in daily use in many regions (Owen, 1994).

Imports from various countries are considered of high or low quality, depending on the country-of-origin. Australia, USA, Japan and Europe scored equally high for quality. India and the Philippines received the lowest scores.

Twenty-two Western-style flavours were rated for suitability for use in three types of food: confectionery, ice cream, and biscuits (cookies). The flavours were considered appropriate for some foods but less so for others: for example, ginger was assessed as very appropriate for confectionery but not for ice cream or biscuits. Chocolate scored as highly appropriate for all three foods. Watermelon and papaya were considered inappropriate for all three. The effect of familiarity on the appropriateness score was measured. It appears to be unnecessary to attempt to sell products made from exotic ingredients not readily available to Western manufacturers. For example, chocolate, strawberry and lemon are highly acceptable flavours at the top end of the market.

CONCLUSION

This data should help companies prepare their market entry or expansion strategies. The study concurs with the advice that adequate knowledge of the target market be obtained, and that products be tailored for it, particularly heeding the importance of taste in purchase decisions. Sensory evaluation of potential export products in the appropriate regions of Indonesia is highly recommended.

REFERENCES

Aminah, A. (1995) Consumer preferences in fruit juice formulation. *Food Agenda 21st Century, 5th ASEAN Food Conference*, Kuala Lumpur, 26–29 July, 165–170.

Bertino, M. and Beauchamp, G.K. (1986) Increasing dietary salt alters salt taste preference. *Physiol. Behav.*, 38, 203–213.

Bertino, M., Beauchamp, G.K. and Jen, K.C. (1983) Rated taste perception in two cultural groups. *Chem. Senses*, 11, 229–241.

Bijlmer J. (1992) Food hawkers in urban Indonesia: The case of Bogor. *4th ASEAN Food Conference*, Jakarta, 17–21 Feb.

Datamonitor, (1997) The leading 50 Asia-Pacific Report: Middle class drives western influences. *Prepared Foods*, July, 35–37.

Druz, L.L. and Baldwin, R.E. (1982) Taste thresholds and hedonic responses of panels representing three nationalities. *J. Food Sci.*, 47, 561–563, 569.

Easton, K., Bell, G.A. and Ng, F. (1997) *Exporting Food to Indonesia: A Guide for Australian Small to Medium Enterprises*. CSIRO and RIRDC, Sydney.

FAS Online Indonesia Food Market Overview, URL: http://ffas.usda.gov/agexport/overviews/indones.html

Hubeis, A.V.S. and Marlin. (1992) Social engineering on street food. *4th ASEAN Food Conference*, Jakarta, 17–21 Feb.

Johansson, B., Drake, B., Pangborn, R.-M., Barylko-Pikielna, N. and Koster, E. (1973) Difference taste thresholds for sodium chloride among young adults: An interlaboratory study. *J. Food Sci.*, 38, 524–527.

Laing, D.G., Prescott, J., Bell, G.A., Gillmore, R., Allen, S., Best, D.J., Yoshida, M., Yamazaki, K. and Ishii-Mathews, R. (1994) Responses of Japanese and Australians to sweetness in the context of different foods. *J. Sens. Stud.*, 9(2), 131–155.

Lundgren, B., Jonsson, B., Pangborn, R.-M., Sontag, A.M., Barylko-Pikielna, N., Pietrzak, E., dos Santos Garruti, R., Moraes, M.A.C. and Yoshida, M. (1978) Taste discriminations vs hedonic response to sucrose in coffee beverage: An interlaboratory study. *Chem. Sens. Flav.*, 3, 249–265.

Lundgren, B., Pangborn, R-M., Barylko-Pikielna, N. and Daget, N. (1976) Difference taste thresholds for sucrose in water and in orange juice: An interlaboratory study. *Chem. Sens. Flav.*, 2, 157–176.

Moskowitz, H.R. (1971) The sweetness and pleasantness of sugar. *Am. J. Psychol.*, 84, 387–405.

Moskowitz, H.R., Kluter, R.A., Westerling, J. and Jacobs, H.L. (1974) Sugar sweetness and pleasantness: evidence for different psychological laws. *Science*, 184, 583–585.

Moskowitz, H.R., Kumariah, V., Sharma, K.N., Jacobs, H.L. and Sharma, S.D. (1975) Cross-cultural differences in simple taste preferences. *Science*, 190, 1217–1218.

Owen, S. (1994) *Indonesian Regional Food and Cookery*. Doubleday, London.

Pliner, P. and Pelchat, M.L. (1991) Neophobia in humans and the special status of foods of animal origin. *Appetite*, 16, 205–218.

Prescott, J., Bell, G.A., Gillmore, R., Yoshida, M., Laing, D.G., Allen, S. and Yamazaki, K. (1993) Responses of Japanese and Australian panel to saltiness in the context of foods. *Chem. Sens.*, 18(5), 616.

Prescott, J., Bell, G.A., Gillmore, R., Yoshida, M., O'Sullivan, M., Korac, S., Allen, S. and Yamazaki, K. (1997) Cross-cultural comparisons of Japanese and Australian responses to manipulations of sweetness in foods. *Food Qual. Pref.*, 8(1), 45–55.

Prescott, J., Bell, G.A., Gilmore, R., Yoshida, M., O'Sullivan, M., Korac, S., Allen, S. and Yamazaki, K. (1998) Cross-cultural comparisons of Japanese and Australian responses to manipulations of sourness, saltiness and bitterness in foods. *Food Qual. Pref.*, 9(1–2), 53–66.

Prescott, J. and Khu, B. (1995) Changes in preference for saltiness within soup as a function of exposure. *Appetite*, 24, 302.

Rozin, P. and Vollmecke, T.A. (1986) Food likes and dislikes. *Annu. Rev. Nutr.*, 6, 433–456.

Slamet, A., (1993) The processed food market in Indonesia. *Food Aust.*, 45(1), 33–34.

Tyler, L. (1998) Asia Beyond 2000: Reviewing trends in tastes and eating habits. *World Ingred.*, March/April, 48–54.

Winarno, F.G. (1992) Food safety, standard and regulations. *4th ASEAN Food Conference*, Jakarta, Feb 17–21.

Yamaguchi, S., Kimura, M. and Ishii, R. (1988) Comparison of Japanese and American taste thresholds. *Proc. 22nd Japanese Symposium on Taste and Smell*, 19.

10

THE ROLE OF STRATEGIC AND APPLIED SENSORY RESEARCH IN THE PERFUME AND FLAVOUR INDUSTRIES

J. LAMBETH

INTRODUCTION

When I was preparing this chapter, I thought, 'Oh dear, I'm going to be writing about chemicals'. If you think that car exhaust emissions are a problem — don't worry, they pale into insignificance once you start talking about chemicals. If one believes the men and women in the street, the majority of the worlds' problems are caused by wars and chemicals. If only we could rid the world of all wars and chemicals — instant Utopia! If you point out that water is a chemical, and that the human body is composed of masses of chemicals, they think you're being frivolous, and not taking the subject of chemicals seriously.

On the other hand, there is 'natural' — AHHH! Public perception is that natural = good for you. 'What about yellowcake', I say, 'That's natural.' — oh, I'm being frivolous again. I point out that 1,8-cineol, the main component of our beloved eucalyptus oil, used for everything from cough lozenges to wool wash products, is so toxic to humans that around five grams is enough to kill an adult human. Shock, horror! I'm sure that after thinking about this revelation, they decide that I'm stretching the truth a great deal. After all, koalas eat eucalyptus leaves all day, in fact, they stink of eucalyptus. So if an 8 kg koala can eat all that eucalyptus, then a human should be able to consume around half a kilo of eucalyptus oil. Reminds me of the scientist who trained a flea to jump when he said 'jump'. After cutting the flea's back legs off, he said 'jump', but the flea didn't move, so he concluded that removing the back legs of fleas make them deaf.

Another problematic perception of chemicals in general, and perfume and flavour chemicals in particular, is that they are cheap, whilst natural products are expensive. I point out that natural orange oil or clove cost only a few dollars per kilo, whilst some of the pyrazines and thiazoles we use cost thousands of dollars per kilo. I'm afraid, however, that a few voices of explanation are not enough. The industries that use chemicals, and the scientific communities that develop them, need to educate the public to appreciate the uses of chemicals in everyday life. The public must learn to use and not abuse chemicals, and to differentiate between the chemicals that are dangerous and toxic to humans and the environment, and those that are not. I'm sure such education can be achieved — starting, I believe, in our schools — even if it takes a decade.

Look what the tobacco industry has achieved. They have convinced a great proportion of the population to consume a highly toxic product and to pay for the privilege of doing so!

SPINNING CONE COLUMN AND CO_2

An area of applied research that has benefited the perfume and flavour industry of late is that of extraction. Extraction techniques to remove the odour and taste principals from botanical material, have remained largely unchanged since their inception over a century ago. Petroleum ether, hexane and benzene have been the traditional solvents, and in theory should produce true-to-type extracts. In practice, even the low levels of heat needed to remove the solvents are responsible for some delicate chemicals breaking down, thereby shifting the odour/taste of the extract to a measurable degree away from the original starting material. During the last decade, superb products have been made available to our industry from two totally different extraction systems. The first is the Spinning Cone Column, an Australian invention hailing from the CSIRO Food Research Division.

The Spinning Cone Column finds its greatest use in the food flavour industry. A large dairy corporation in New Zealand is using Spinning Cone Columns to remove the grassy taste from milk. One of the great advantages of the Spinning Cone Column is that it is a continuous extractor, rather than a conventional batch extractor: a tomato extract produced by this method was a 12 000 times concentrate, but exhibited no harsh or burnt notes, being remarkably true to the fruit.

The other great advance in extraction technology is the use of carbon dioxide (both sub- and super-critical), again, at this stage, mainly for flavour extracts. The advantages of using liquid CO_2 is that the extraction of the botanical material takes place at such reduced temperatures that degradation of constituents by heat is eliminated. The resultant extracts are truly remarkable. Apple extracts that really do smell like a freshly cut apple! A rosemary extract, which for the first time in my 30 year career, actually smells like the rosemary plant when you crush the leaves, rather than camphor and eucalyptus, with passing resemblance to rosemary in the background. The bottom line is that natural flavours and flavour compounds are becoming better and tastier, with consumers being the winners.

HEADSPACE ANALYSIS OF LIVING BOTANICAL MATERIAL

An exciting area of applied sensory research that is beginning to yield interesting results for the fragrance and flavour industry is that of headspace analysis of living botanical material. This process involves closed-loop stripping of volatiles emanating from fruit or flowers, and is carried out by means of a flask placed over, for instance, a flower either growing in a field or container. A small pump is used to circulate air across the flower and through a carbon filter, which traps the odour vapours. After a given time, usually many hours, the carbon filter is removed and the entrapped odour is put through a gas chromatograph and mass spectrometer, to analyse and identify the components.

Extremely interesting and unexpected results have been elucidated by this research. Firstly, unstable chemicals with a very short life span have been identified. These chemicals have not been found in conventional, experimental or commercial extracts, as they have been lost or destroyed during processing. Many of these chemicals, however, have proved to be missing links in the odour/taste profiles of natural extracts. The fact that they are unstable means that they are of limited commercial use. However, by examining their chemical structure, it is sometimes possible to find a close relative that is more stable. Another alternative involves making an odour/taste analysis of the component, then trying to find as close as possible an odour/taste substitute regardless of chemical similarity.

The structure of the extract is also of great potential interest to perfumers and flavourists. Those of you who have had contact with essential oils, extracts and absolutes have probably noticed that they smell heavier, richer and less fresh and natural than the botanical material from which they are derived. For instance, if you smell a jasmine flower, and jasmin absolute, you notice a distinct difference. We now find this is largely to do with the proportion of top note versus middle note versus base note. When you smell a live jasmine flower, you are smelling mainly the top notes, a small proportion of middle notes, and even less base notes.

The jasmin absolute, however, reveals mostly base notes, some middle notes and less top notes. This is the inverse of the natural flower. By applying this knowledge, perfumers and flavourists are now realising that re-proportioning formulations with a bias towards top notes will yield more 'true to life' fragrances and flavours. Dragoco have applied this knowledge to produce a range of floral bases under the *Aurascent* umbrella. These bases are used by Dragoco's perfumers to obtain what could be coined as 'a natural advantage' in fragrance creations.

BIODEGRADABILITY — A STORY OF CONTRADICTIONS

The manufacturer's desire is to have, for instance, a washing powder perfumed at 0.2%, which is stable enough to withstand the concentrated surfactants, bleaching agents, etc., contained in the market product. So when the consumer purchases the product six weeks or six months later, the product, when opened, smells fresh and inviting.

When our customer uses the powder in the wash, a pleasant smell is expected. After vigorous washing and rinsing a couple of times, our customer still expects a fresh, reassuring smell whilst hanging out the washing. A sun-dried smell, but with a hint of fragrance, is expected when the clothes are brought in after a day in the summer sun. Our customer expects the tedium of ironing to be (slightly) relieved by a hint of fragrance as the scorching iron is passed over the garments. A few days later, when the ironed article is removed from the linen cupboard, a few lingering fragrance molecules stagger out to greet our customer — after being soaked in concentrated surfactants for six months, dissolved and beaten in water, left in the blazing sun and scorched with an iron.

Maybe you think I'm exaggerating, but about ten years ago a product manager from a detergent manufacturer confided in me that they would like to change their current perfume because it was lasting too long. A few mothers had written in to say that their male children had been taunted at school because they had a 'sissy' smell.

Applied science has given us these super stable chemicals. Chemicals capable of withstanding what 50 years ago was regarded as an impossible ask. All detergent perfumes now contain such 'cast iron' stable chemicals as a standard feature, and therein lies the problem. What is welcomed on the clothes is not welcomed in the environment, now that science has enabled us to detect chemicals in the environment at parts per trillion. These super stable chemicals, traces of which did not attach themselves to the clothes during the washing process, but were discarded with the waste water, are showing up, good and stable, just as they were designed to be. Some have also found their way back into the food chain, and have been detected in breast milk. No deleterious health risks have been found, but environmentalists are concerned just by the fact that they are there. I'm sure this is not the end of the story, either. More and more chemicals are going to be discovered polluting our environment and our bodies.

THE FUTURE

Last week I received a newsletter from the European Flavour and Fragrance Association expressing concern for the growing consumer-driven demand to ban nitromusks. Previously, all suspect fragrance chemicals have been deleted or regulated by RIFM/IFRA (Research Institute for Fragrance Materials/International Fragrance Research Association), that is, from the top down. This new to RIFM/IFRA phenomenon of the consumer dictating deletion of chemicals has come as a nasty shock, but I suspect it is indicative of future trends — namely consumers rejecting products on the basis of information presented by the media, be it correct or incorrect.

This may lead to a situation where only natural, or natural and nature-identical ingredients, will be deemed acceptable to the media/consumers. Such a policy would result in virtually all fragrances currently in use being unacceptable. The consequences of such a policy to our industry would be catastrophic. What is the answer? Can scientists help here? Sure, we are never satisfied. We asked for stable chemicals, and scientists gave them to us. Now we want scientists to get rid of them once they have completed their task, and before they pollute our environment. Seems like a tough ask to me, but then again, I'm not a scientist.

As an analogy, the case of plastic containers in general, and plastic bags in particular, come to mind:

Requirement: Good strong plastic bag to carry home groceries. But on disposal, bag ends up polluting land fill.

Answer: Use a plastic that breaks down over time with the influence of sunlight.

In the case of our super stable laundry detergent chemicals, the answer is unfortunately not so simple. If the sunlight solution were to be used, we would have no residual fragrance retention on the dried washing, which would be an unfortunate compromise.

Perhaps we need to look carefully at the mechanism whereby nature degrades its essential oils in fallen botanical material. After all, natural essential oils are composed of the full gamut of chemical groups. This includes very substantive chemicals such as ambrettolide (a musk chemical occurring naturally in ambrette seed oil, from a variety of hibiscus); and very stable chemicals such as cineol, from our eucalyptus species, which is stable enough to withstand chlorine bleach. Nature is able to dispose of all these materials after use.

USING ANALYTICAL SENSORY TECHNIQUES TO UNDERSTAND WINE PREFERENCE

A.C. NOBLE

INTRODUCTION

Sensory evaluation is a scientific discipline used to quantify, analyse and interpret reactions to the sensory properties of wine or any product. Traditionally it is assumed that the primary sensory characteristics that affect our response to wine are appearance and flavour. However, perception of flavour involves integration of the separate sensations of smell, taste, and touch. It is influenced by appearance, as well as by reputation of the winery or wine region, experience and expectation of the consumer, price, and appearance of the package. In analytical sensory tests, the effects of all extraneous clues that bias perception of the aroma and taste of wine are removed. Thus, studies conducted using these analytical methods can yield valuable quantitative information about the effects of climate, soil, vineyard treatments or winemaking processes on wine flavour.

Although analytical (objective) sensory analysis provides precise information, it cannot predict consumer acceptance or wine quality. Examples of descriptive analysis of wines stored at elevated temperatures, and of wines made from different vineyard locations, are presented. To interpret the effect of treatments on consumer acceptance, techniques for relating descriptive data to marketing information and consumer testing are briefly introduced.

INTRODUCTION TO DESCRIPTIVE ANALYSIS (DA)

Descriptive analysis provides quantitative, analytical information about wine flavour. Precise terms are selected to describe the important aroma notes, flavours, and tastes that differ among the samples. Terms such as those listed on the Wine Aroma Wheel can be defined by preparation of reference standards as described previously (Noble *et al.*, 1987; Noble, 1988). In most cases, reference standards are prepared to define the aroma terms using a neutral base wine, spiked with a food product, such as a slice of bell pepper or drop of vanilla extract. Judges are then trained in the use of rating scales and in consistent use of the terms, prior to rating the intensities of these attributes in the experimental wines. For more details about conducting descriptive analysis, see Heymann *et al.* (1993), and Lawless and Heymann (1998).

The type of information provided by descriptive analysis is illustrated by a study of Chardonnay wines which had been stored at 40°C for 0 to 45 days, to simulate the effects of elevated temperatures during wine transport or storage. Trained judges rated attributes describing the aromas of the wines. The average intensity ratings for these nine terms are plotted in Figure 11.1. In this polar co-ordinate graph, the centre of the figure represents low intensity with the distance from the centre to the intensity rating corresponding to the relative intensity of each wine for the attribute. By connecting the mean ratings, the profiles of wines are shown. Heating the wines for 15 days produced a significant decrease in fruity and floral aromas. Storage for 30 days resulted in a dramatic reduction in the fruity characters, and produced a corresponding increase in terms such as tea/tobacco and honey, which are associated with aging or possibly older or oxidised wines (de la Presa Owens and Noble, 1997).

One cannot, however, draw conclusions from just a few wines, in studies of the effect of vineyard location, climate or soils. Meaningful results will only be obtained from larger data sets. To explore the effect of vineyard location and soils, a descriptive analysis of 19 Napa

Valley Cabernet Sauvignon wines was conducted (Spears, 1990) over three years. Clearly, descriptive analysis data for 19 wines cannot be presented or interpreted by the simple cobwebs in Figure 11.1. To simplify the interpretation of large data sets, principal component analysis (PCA) is often applied to descriptive analysis of wine flavour, to show relationships among the attributes and among the wines. A first principal component (PC) is derived to explain the maximum amount of variance in the data. The second PC is then extracted in the same way, from the remaining unexplained variance. Each principal component is a linear combination of the variables (descriptive attributes). Thus, plotting the loading for each term as a vector instantly allows one to see the important differences in sensory properties of the wines.

In Figure 11.2, the wine factor scores and attribute loadings (plotted as vectors) are shown for the first two principal components from the PCA of mean intensity data for one vintage of these Cabernet Sauvignon wines. The first principal component (PC 1) is primarily a contrast between vegetative notes versus berry, vanilla and butterscotch, terms which are more intense in wines such as number 12. Wines located on the left, such as wine 13, are low in the fruity notes and high in vegetative ones. Interestingly, there is no clustering of wines by region of origin, suggesting that factors other

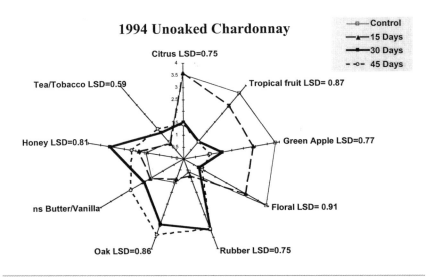

Figure 11.1
Mean intensity ratings for Chardonnay wines stored at 40°C for 0, 15, 30 and 45 days. (n = 14 judges × 3 reps) (de la Presa Owens and Noble, 1997)

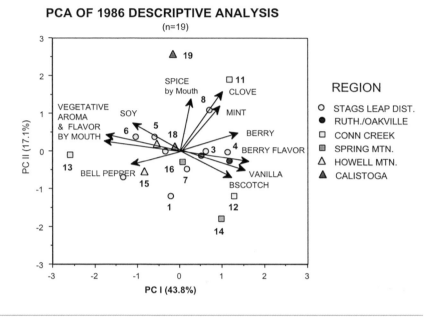

Figure 11.2
Principal Component Analysis of Descriptive Analysis of 19 Cabernet Sauvignon wines.
Factor scores shown as points, coded by region of origin; attribute loadings shown as
arrows for principal components (PC) I and II (n = 18 judges × 2 reps). (Spears, 1990)

than the location of the vineyard have stronger influence on wine
flavour. In fact, wines 12 and 13 were both produced by the same
winery from vineyards separated by 1 km, yet they show very large
differences in flavour.

To relate the vineyard and winemaking factors to these differences
in flavour, Partial Least Squares analysis (PLS) was used. This statistical method models the variability in two data sets. One set of data
included soil data for each vineyard site, while the other data set was
provided by descriptive analysis of the wines. From this analysis, an
association was found between wines with higher intensity of vegetative aroma and flavour by mouth and vines grown on soils with
higher water holding capacity. Conversely, fruitier wines, high in
berry aroma and flavour, were associated with older, gravelly soils
with poor water holding capacity. Consistent with this model, the
soil at site 12 is shallow, sandy, and nutrient-poor with a low water
holding capacity, and wine 12 is quite fruity. In contrast, the vegetative 13 wine was produced from a deep, clay-rich, older soil, high in
nutrients and water holding capacity (Elliott-Fisk and Noble, 1992).

CONSUMER ACCEPTANCE OF WINES

In these two examples no reference to 'wine quality' has been made. Preference *and* quality are composite subjective responses to the sensory properties of wine, based on one's expectations for a given wine type, which are a function of one's previous experience with wines. Quality is an individual response, varying from person to person because of different sensitivities, expectations and preferences. The most experienced, skilled and sensitive wine judges may often have differences of opinion about wine quality, and also may be unable to predict consumer acceptance. For example, it is a strongly held opinion in the wine industry that storage of wines at elevated temperatures has a very deleterious effect on wine quality. However, storage of wines has been shown to increase certain varietal aromas as volatiles are released from glycosidic precursors (Francis *et al.*, 1992). Although the differences in heated wines were precisely described in Figure 11.1, no preference evaluations were conducted. Therefore it is not known whether quality was lowered systematically by heating, or if an initial improvement of quality occurred upon accelerated aging for 15 days.

Most winemakers will insist that they do *not* like or want to make Cabernet Sauvignon, Merlot or Sauvignon blanc wines with intense bell pepper (vegetative) aromas. In fact, many assume that a wine intense in bell pepper aroma will be disliked by most consumers. Their reasoning is twofold: firstly, they assume that it is safe to generalise from their own preferences to those of consumers; secondly, they assume that a more complex wine is 'better', and clearly one dominated by intense vegetative aroma is not complex. Generally their assumption is true, but it is not always the case. For example, the scores for dry Sauvignon blanc wines at a wine show were higher for wines with higher levels of methoxy isobutylpyrazine (MIP), a compound which elicits a strong bell pepper aroma (Allen *et al.*, 1988). Although there was a wide spread in wine scores when plotted against the MIP level, the two highest scoring wines had over 30 ppt of MIP.

Unfortunately, consumer preference cannot be predicted from opinions of 'experts', any more than it can be predicted from the concentration of a single compound. Not only do experts disagree among themselves, but consumers represent a diverse population and have very different acceptance standards. Consumer preference is a function of individual expectations, previous experience with a wide variety of wines, innate likes and dislikes, and sensitivity to aroma and taste compounds. To illustrate this, three different patterns of consumer acceptance could occur in preferences for wines varying in MIP concentration. Some consumers will prefer an intense bell pepper aroma (LIKERS), and their preference increases with the concentration of MIP. Other tasters (DISLIKERS) dislike wines more and more as the bell pepper intensity (and level of MIP) increases. A final pool of wine consumers could be called OPTI-MIZERS: these are the tasters who dislike high levels of MIP, but like best the wines with some obvious bell pepper character produced by concentrations of MIP above threshold levels.

This hypothetical data, of course, doesn't begin to deal with complexities that we see in real wines. In real wines, not only will the MIP level vary, but so will the concentration of masking aromas, such as oak-extracted volatiles. Nor, of course, does it consider the obvious issues other than wine flavour which influence consumer preference. Factors such as image, reputation, label, and price, all greatly drive consumer acceptance. For many years, wineries have known that selling white wines in clear bottles, especially Sauvignon blanc, is courting disaster. Exposure of Sauvignon blanc to direct sunlight often results in production of off-odours in less than five minutes; despite this, many wineries still bottle Sauvignon blanc in clear glass bottles because of concern about appearance rather than the potential effect on aroma.

To identify the factors that influence consumer perception, a variety of multivariate statistical techniques have been proposed. Internal preference mapping is a technique in which preference is rated by target consumers, and a 'product space' is developed from these data by a form of multidimensional scaling (MDS) such as MDPREF (MacFie and Thomson, 1988; Nute et al., 1988; Schlich and McEwan, 1992). Similarly, subjects can be asked to group wines on the basis of their similarity or preference, and the results analysed by another MDS procedure, MDSORT. The advantage of such techniques is that wines located closely in the space, or consistently grouped together, are perceived to be similar. Another advantage is that the consumer makes the judgement with no instructions, and can use or ignore any available information. A disadvantage is that the characteristics of the preferred wines must be determined by external preference mapping methods, which relate consumer acceptance ratings to information such as descriptive analysis data for the same wines. The preference data can be related to analytical sensory (or instrumental) data by techniques such as Partial Least Squares Regression (PLS), Principal Component Regression, Procrustes Analysis (Risvik et al., 1994) or preference mapping programs (see the text edited by Naes and Risvik, 1996). By analysis of demographic data, marketing methods can be tailored for the segments of consumers revealed by the preference mapping study.

Much work remains to be done in probing factors driving consumer preferences. It is commonly accepted that naive consumers focus on taste, preferring sweeter wines, whereas more experienced wine drinkers prefer 'more complex' wines. Yet no one has explored the number of dimensions that different types of wine consumers perceive. What factors influence different consumer groups? Finally, a crucial issue, which has not yet been studied in consumer testing of wines or other products, is to determine what is the effectiveness of momentary preference judgments in predicting purchase decisions.

REFERENCES

Allen, M.S., Lacey, M.J., Harris, R.L. and Brown, W.V. (1988) Sauvignon blanc varietal aroma. *Aust. Grapegrower Winemaker*, 3, 52–56.

de la Presa Owens, C. and Noble, A.C. (1997) Effect of storage at elevated temperatures on aroma of Chardonnay wines. *Am. J. Enol. Vitic.*, 48:3, 310–316.

Elliott-Fisk, D.L. and Noble, A.C. (1992) Diversity of soils and environments in Napa Valley, California and their influence on Cabernet Sauvignon wine flavours. In de Blij, H. (ed), *Viticulture in Geographic Perspective*. Proceedings 1991 Miami AAG Symposium, pp 45–71.

Francis, L., Sefton, M.A. and Williams, P.J. (1992) Sensory descriptive analysis of the aroma of hydrolysed precursor fractions from Semillon, Chardonnay and Sauvignon blanc grape juices. *J. Sci. Food Agric.*, 59, 511–520.

Heymann, H., Holt, D.L. and Cliff, M.A. (1993) Measurement of flavor by sensory descriptive techniques. In Ho, C.–T. and Manley, C.H. (eds), *Flavor Measurement*. Marcel Dekker, NY, pp. 113–132.

Lawless, H.T. and Heymann, H. (1998) Sensory Evaluation of Food: Principles and Practices. Chapman Hall, NY, 819pp.

MacFie, H.J. and Thomson, D.H. (1988) Preference mapping and multidimensional scaling. In Piggott, J. (ed), *Sensory Analysis of Food*. Elsevier, London, pp. 381–409.

Naes, T. and Risvik, E. (1996) (eds) Multvariate analysis of data in sensory science, Elsevier, NY, 348pp.

Noble, A.C., Arnold, R.A., Buechsenstein, J, Leach, E.J., Schmidt, J.O., Stern, P.M. (1987) Modification of a standardized system of wine aroma terminology. *Am. J. Enol. Vitic.*, 38, 143–146.

Noble, A.C. (1988) Analysis of Wine Sensory Properties. In Linskens, H.F. and Jackson, J.F. (eds), *Wine Analysis. Vol. 6. Modern Methods of Plant Analysis*. Springer-Verlag, London, pp. 9–28.

Nute, G., MacFie, H. and Greenhoff, K. (1988) Practical application of preference mapping. In Thomson, D.H. (ed), *Food Acceptability*. Elsevier, Amsterdam, pp. 377–386.

Risvik, E., McEwen, J., Colwill, J., Rogers, R. and Lyon, D. (1994) Projective mapping: a tool for sensory analysis and consumer research. *Food Qual. Pref.*, 5, 263–269.

Schlich, P. and McEwan, J. (1992) Preference mapping. A statistical tool for the food industry. *Sci. des Aliments*, 12, 339–355.

Spears, T. (1990) Evaluation of the effects of soil and other geographic parameters on the composition and flavor of Cabernet Sauvignon wines from the Napa Valley. M.Sc. Thesis. Univ. California, Davis.

ACCOUNTING FOR SEVERAL RELATED SOURCES OF VARIATION IN CHEMOSENSORY PSYCHOPHYSICS

J.C. WALKER, M. KENDAL-REED AND W.T. MORGAN

INTRODUCTION

A major issue in the collection and interpretation of human chemosensory psychophysical data is *variation*. The ability to reach conclusions in the academic or industrial laboratory often hinges on the confidence that the experimenter has in the reliability of a finding that two stimulus conditions are rated as different. We use the term 'stimulus condition' to cover anything from a well-controlled presentation of a single odorant or tastant chemical, to a product as complex as a food, beverage or cigarette. In practice, decisions as to whether apparent changes in sensory magnitude or quality are 'real' mean that some sort of judgement is made of the amount of variation within the set of responses to each stimulus condition. This is then compared to the difference between the average responses to the two or more stimulus conditions.

Researchers, particularly in the industrial setting, are often forced to estimate the likelihood that a generally similar result would be obtained under different circumstances. For example, the researcher may speculate what further information would have been obtained had some combination of the following been the case: larger numbers of participants, a greater amount of data collected from each participant, or more precise control over stimulus composition or presentation.

Typically, a conceptual model of the inter-locking or 'nested' sources of variation, separate from the stimulus conditions to be compared, is not envisioned. Such a model would contain the following elements:

1 Stimulus variation The physical or chemical variation among different presentations of what are treated as identical stimuli;

2 Intra-session uncertainty The variation for a given individual in the perception of identical stimuli presented repeatedly in the same test session;

3 Intra-individual (between session) The true variation for a given individual in the perception of identical stimuli over the course of days or weeks; and

4 Inter-individual variability The true variation in sensory perception among individuals in the sample and, by inference, those in the target population.

The importance of differentiating among these sources and understanding the relative impact of each derives from the fact that apparent variation at one level can actually be a manifestation of variation at a lower level. Cain (1977) provided an instructive example on this point by showing that the presumed poor ability of the olfactory system to detect differences in intensity was largely attributable to inadequate control over stimulus parameters. When stimulus 'noise' was dealt with effectively, it was shown that the intensity discrimination capability of the olfactory system (DI/I threshold of ~ 0.05) was comparable to that of the visual and auditory systems.

It is the intent of this paper to describe an approach to clarifying and quantifying sources of variation in odour sensitivity and magnitude, so that the basis for decision-making is improved. Specifically we:

1 describe an approach for processing raw sensory ratings in order to estimate the stimulus intensities (termed iso-response concentrations) required for several perceptual magnitudes, for a given participant-by-session combination;

2 describe a method for estimating the uncertainty surrounding each of the iso-response concentrations calculated for a given session-by-participant combination;

3 develop a procedure for estimating the proportion of apparent intra-individual variation that is accounted for by intra-session uncertainty;

4 describe a procedure for estimating inter-individual variation in a sample of participants and, by inference, a population; and

5 suggest some practical advantages and applications of improved understanding of variation sources.

THE SAMPLE DATA SET

Our approach has been to start with stringent control over stimulus parameters, and then conduct the research in ways that allow each of the variation sources to be evaluated explicitly. We illustrate our approach using odour psychophysical ratings from a recent study (Kendal-Reed et al., 1998). Thirty-one participants completed four test sessions, in each of which were presented 10 trials at each of four propionic acid concentrations, and ten air control trials. Propionic acid concentrations were generated by a computerised air-dilution olfactometer, which produced verifiably accurate odorant concentrations. Each 15-second stimulus presentation, to a modified oxygen mask fitted with a pressurised cuff (to ensure a snug fit), began with the onset of an inhalation.

At the end of each trial, participants reported odour magnitude using a mouse and a computerised visual analogue scale. Participants were instructed to mark the leftmost end of the scale (coded as 0) to indicate an absence of odour, and to rate the magnitude of any odour sensation relative to the strongest odour sensation experienced prior to the study. This 'previous high' point was labelled on the scale and was coded as 65. The range of values from this point to the extreme right end of the scale (coded as 99) allowed the participant to report magnitudes that exceeded those encountered before the study.

The sample data set therefore consisted of 6200 odour magnitude ratings (31 participants x 4 sessions/participant x 50 ratings/session). Figure 12.1 depicts these data. Each plot reflects some unknown contribution from each of the sources of variation listed above.

ALLOCATION OF VARIATION TO SOURCES

The first step in our analysis was the development of a procedure for expressing the strength of odour perceptions in units of odorant concentration. Our approach, termed iso-response estimation, is illustrated in Figure 12.2. For each session-by-participant combination, the concentrations (in log ppm) corresponding to 10, 25, 50, and 75% of the 'distance' from the mean response on clean air trials up to the maximum rating are calculated. In this illustration the mean of the participant's odour strength ratings on clean air trials was 10, and the highest rating for this session was 95, yielding a 'distance' of 85. Concentrations corresponding to the 10, 25, 50 and 75% ratings are determined by simple interpolation: 18.5 (=10 + 10% of 85), 31.25 (=10 + 25% of 85), 52.5 (=10 + 50% of 85), and 73.75 (=10 + 75% of 85). Note that these points are not percentile values. They simply denote various points along a continuum, from the mean rating on blank trials up to the highest rating, for a given session. These stimulus values are termed iso-response concentrations, since they approximated the stimulus intensities at which various sensation magnitudes are equated for different participants. This approach is analogous to the determination, for a set of sugars, of the concentration of each which corresponds to a given sweetness intensity.

Since each session yielded a single estimate of the concentration required for each of the four perceptual magnitudes, there was no direct way to determine the degree to which variation over the course of a session might cause these concentration estimates to vary. For estimating the uncertainty associated with each of the four iso-response concentrations calculated for a given participant-by-session combination, we used a bootstrap re-sampling technique (Efron, 1982). This is a form of simulation that samples from the ratings obtained during an actual testing session, but makes no assumptions about the statistical distributions of these data. Thus, this bootstrap approach approximates the underlying distributions without specifying them, so that the distributions of derived statistics (like the iso-response concentrations described here) can be 'sampled' and summarised. One re-sampling simulation run consisted of drawing 10 samples (with replacement) from the 10 observed ratings for each of

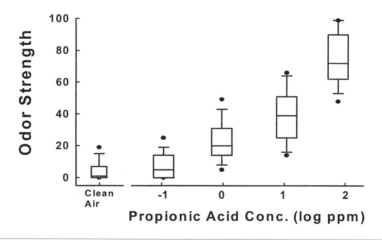

Figure 12.1

Box-whisker plots showing the distribution of the total of 1240 odour magnitude ratings made by 31 participants, for each of the five stimulus conditions. The 5th, 10th, 25th, 50th (median), 75th, 90th and 95th percentiles are represented by, respectively, the bottom filled circle, the lower tail, the lower box edge, the line passing through the box, the top box edge, the top tail and the top filled circle.

the four propionic acid concentrations, and for the clean air trials. Based on these 50 simulated odour strength ratings, iso-response concentrations were determined as described above (see Figure 12.2). Intra-session uncertainty, for each perceptual magnitude, was then defined as the variance of the results of 100 bootstrap simulation runs. For a given participant, the four within-session variance estimates (one for each session) were assumed to be independent estimates of the same quantity and were thus pooled (by averaging) to provide an overall estimate of the within-session variance component.

For each combination of participant and perceptual magnitude, the apparent between-session variation was expressed in terms of the variance of the four iso-response concentrations (one per session). Since both within- and between-session variation were in the same units, variance components analysis could be used to determine the 'pure' between-session variance (Box et al., 1978). With a nested or hierarchical design, as we have employed in this work, estimates of sample variance at different levels of summarisation include all 'lower' level variance components. The 'pure' or adjusted between-session variance is therefore obtained by subtracting the within-session from the apparent between-session variance.

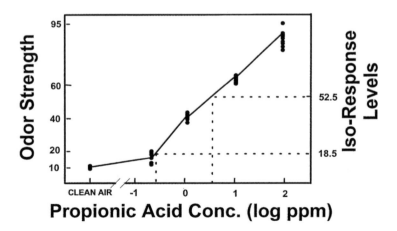

Figure 12.2

Illustration of method used to summarise the results from each participant-by-session combination. The concentrations corresponding to various perceptual levels were determined by interpolation from a plot of the sensory ratings compared with the four propionic acid concentrations. In this example from a typical normal participant, concentrations of $10^{-0.5}$ corresponds to the 10% iso-response point and a concentration of $10^{0.5}$ correspond to the 10% and 50% iso-response point, respectively. (Kendal-Reed, M., et al., 1998)

Figure 12.3 summarises the results of this approach, for each of the 31 participants. There is a decline in variation with increasing perceptual magnitude. Roughly 70% of the participants show little variation over time. This pattern of results provides strong evidence that, when the requisite care is taken with both stimulus precision and the numbers of trials per stimulus condition, true fluctuations in odour sensitivity are typically quite small. For the remaining participants, both the apparent and adjusted between-session variances demonstrate some instability or inconsistency over time. The advantage of our adjusted between-session variance estimates, for assessing mean levels of between-session variation or for comparing individuals, is most apparent for the lowest two iso-response values (10% and 25%). In the absence of our adjustment, variation over time would be over-estimated to some unknown degree, and the experimenter would make some errors in ranking participants according to consistency of responses.

While the use of variances does make it possible to apportion sources of variation, the values must be converted to standard deviation units in order to express variation in terms of distances on a logarithmic scale of odorant concentration. (Note that, as illustrated in Figure 12.2, the iso-response determinations were made by interpolation from a plot of odour strength compared with log ppm). Consider the highest adjusted between-session variance of 0.84, which was exhibited by participant 16 at the 10% iso-response level. The square root of this value (0.92) provides the adjusted between-session standard deviation in terms of log ppm. Thus, a logarithmic distance of 1.84 ($10^{1.84}$ = 69-fold) would be expected to encompass the center 68% (two standard deviation units) of this individual's 10% iso-response values, if a sufficiently large number of trials were included each session. As a progressively larger number of sample trials were included per session, within-session uncertainty would approach zero. This would cause apparent between-session variation to fall toward adjusted variation.

INTER-INDIVIDUAL VARIATION

Having dealt with between-session variation within individuals, the possibility of extending the approach to a treatment of between-participant variation was considered. However, given the marked differences among participants in terms of true between-session variation, no attempt was made to determine what proportion of apparent between-participant variation might be attributable to between-session variation. Instead, the iso-response concentration

Figure 12.3
Depiction of the apparent, and adjusted, between-session variances, separately for each combination of participant and perceptual magnitude. Vertical lines are added to aid comparisons among the four magnitudes.

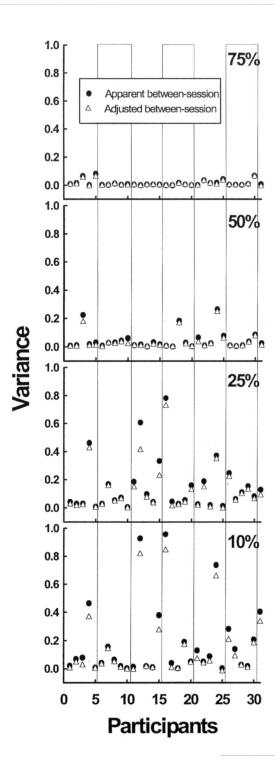

values were simply averaged across sessions, separately by perceptual magnitude, for each participant. In this way, the performance of each individual was summarised by a set of four values. Then the mean, standard deviation and variance were calculated for each of the four perceptual magnitudes. These are shown in Table 12.1.

Table 12.1
Statistical summary of variation among 31 participants

Perceptual magnitude	Mean (log ppm)	Standard deviation	Variance
10%	-0.56	0.34	0.11
25%	0.15	0.34	0.12
50%	0.97	0.27	0.07
75%	1.52	0.18	0.03

Note: Means are in units of log ppm and standard deviations are therefore expressed in terms of logarithmic distances on a ppm scale.

POSSIBLE APPLICATIONS

We suggest that our approach for the processing of sensory data could be of use in the following ways:

1   This approach may be useful in the analysis of existing data sets, in order to better understand the pattern of responses of groups of individuals to a set of products or other stimulus conditions. For example, estimating the relative importance of various nested sources of variation along the lines we suggest could help one understand the reasons for 'inconsistency' within or among different individuals exposed to the same stimulus condition.

2   Preliminary work using the approach we outline, prior to much more extensive testing, would likely provide valuable guidance for designing optimally informative experiments. Suppose that a researcher determines from such preliminary work that, relative to true between-session variation within individuals, both intra-session and inter-individual variation are likely to be quite small. In this case the design for the larger study could incorporate a relatively modest number of participants, each of whom would be tested repeatedly over the course of several sessions.

3   Finally, our methods for quantifying different sources of variation could be used to minimise variation in responses within each stimulus condition by selecting, for detailed testing, those individuals exhibiting the greatest sensory 'precision'.

ACKNOWLEDGMENTS

Research described in this paper was supported by the Center for Indoor Air Research, the RJR-UNC Collaborative Olfactory Research Program and NIH Grant DE06957 (Prof. D.W. Warren, PI). Figure 12.2 was reproduced with permission from Kendal-Reed, M., *et al.*, 1998.

REFERENCES

Box, G.E.P., Hunter, W.G. and Hunter, J.S. (1978) *Statistics for Experimenters.* John Wiley and Sons, NY, NY, pp. 571–583.

Cain, W.S. (1977) Differential sensitivity for smell: 'noise' at the nose. *Science,* 195, 796–798.

Efron, B. (1982) The bootstrap. In Efron, B. (ed), *The Jackknife, the Bootstrap and Other Resampling Plans.* Society for Industrial and Applied Mathematics, Philadelphia, Pa., pp. 29–35.

Kendal-Reed, M., Walker, J.C., Morgan, W.T., LaMachio, M. and Lutz, R.W. (1998) Human responses to propionic acid. I. Quantification of within- and between-participant variation in perception by normal and anosmic participants. *Chem. Senses,* 23, 71–82.

PRODUCT MAPS FOR SENSORY RANKING AND CATEGORICAL DATA

D.J. Best, J.C.W. Rayner, and M. O'Sullivan

INTRODUCTION

Preference ranking and category scaling are well known methods for product comparison. Statistical procedures are commonly used to verify that apparent differences between products are due to other than chance effects. If scores are assigned to the categories, then analysis of variance is often the statistical procedure employed. If only rankings are available, or if a nonparametric statistical method is used, then the Kruskal-Wallis test for one-way data and the Friedman test for two-way data are available (Conover, 1998). In this paper we complement these nonparametric methods with product maps, which visually summarise the differences between the products.

In the following we will consider consumer sensory trials. Commonly, continuous line scales are used for consumer trials, but such scales are inherited from trained laboratory taste test panels. Untrained consumers may use the scales in different ways. Some may spread their line scale scores, while others will use a narrow range. This may invalidate the usual analysis of variance method, and suggests that category scores or ranking would be a safer way to proceed with consumer trials.

ONE-AT-A-TIME CATEGORICAL SCORES

It can be argued that for consumer or market research trials, subjects should evaluate one product only, as this more closely mimics the way consumers receive their food. Suppose this one-at-a-time pro-cedure, or in statistical terminology, a completely randomised design, has been followed, and five products have been rated by 210 consumers using a five-point category scale. The data are summarised in Table 13.1.

Table 13.1
Response frequencies in a consumer taste test

	Response category				
Product	Awful	Not good	Okay	Like	Terrific
A	9	5	9	13	4
B	7	3	10	20	4
C	14	13	6	7	0
D	11	15	3	5	8
E	0	2	10	30	2

The Kruskal-Wallis statistic (adjusted for ties) is 35.44 on four degrees of freedom with P-value less than 0.01, indicating that the mean ranks of the products significantly differ. Most statistics pack-ages will perform this calculation, and Conover (1998) gives details. However, it is often important to look at differences in the spread or dispersion of the scores or ranks, as well as the differences in means that the Kruskal-Wallis statistic picks up. As we discuss below, dif-ferences in dispersion can identify important market segmentation effects. To look at mean and dispersion effects simultaneously, we proceed as follows:

A PRODUCT MAP

Calculate M_i , the linear or mean effect, and V_i , the quadratic or dispersion effect, as outlined in the Appendix. A plot of the (M_i, V_i) for all t products gives a *product map*, and Figure 13.1 does this for the Table 13.1 data.

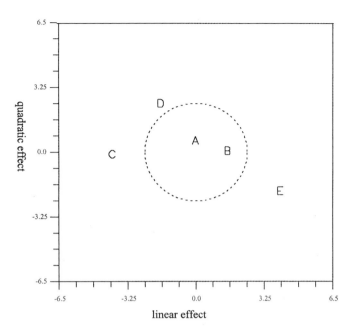

Figure 13.1
Product map and 95% confidence circle for the consumer category scale data in Table 13.1.

The values of the M_i separate the products according to mean scores, and in fact $\sum_{i=1}^{t} M_i^2$ is very similar to the Kruskal-Wallis statistic. A high negative V_i (low dispersion) implies the scores are clumped around the middle categories, whereas a large positive V_i (high dispersion) implies the scores are at one end or are in two clumps around both low and high scores. Two clumps indicate lack of consensus amongst consumers, or market segmentation. One clump indicates consensus. From the data from Table 13.1 we can thus conclude that product E is well liked, while product C is not. This much could have been deduced from a mean scores analysis. However we also see that product D stands out when V_i values are considered, and this emphasises the market segmentation of consumers for this product. This is in agreement with a 'by-eye' inspection of the distribution of scores for D in Table 13.1.

The circle shown in Figure 13.1 is a 95% confidence circle with the equation $x^2 + y^2 = 6$. Values of (M_i, V_i) outside this circle are significantly different to what would be expected if there were no product differences. Another discussion of the Table 13.1 data, and a reference to its source, are given in Hirotsu (1990).

A PRODUCT MAP FOR RANKING DATA

Baba (1994) gave the data in Table 13.2 for 20 consumers' overall preference ranking of eight products. Notice here that each consumer ranks all eight products. In statistical terminology this is a randomised block design. To account for each consumer ranking all eight products, we multiply the M_i and V_i calculated in the Appendix by $\sqrt{(t-1)/t} = \sqrt{7/8} = 0.9354$ here. Also, the radius of the confidence circle is multiplied by this same quantity. Otherwise, we proceed as above after obtaining the frequency data shown in Table 13.3.

Table 13.2
Rankings of products

Judge	Product								Judge	Product							
	A	B	C	D	E	F	G	H		A	B	C	D	E	F	G	H
1	2	5	8	6	3	4	7	1	11	6	7	1	5	4	2	8	3
2	1	5	8	4	6	3	7	2	12	3	1	2	4	6	5	8	7
3	1	5	7	3	6	4	8	2	13	8	1	7	6	5	3	4	2
4	4	2	5	6	1	7	8	3	14	3	1	6	2	4	5	8	7
5	1	4	2	6	3	7	8	5	15	3	1	6	4	2	5	8	7
6	1	7	5	4	2	6	8	3	16	3	1	6	2	5	4	8	7
7	4	1	2	3	6	5	7	8	17	8	5	4	2	1	3	6	7
8	4	1	2	3	5	7	6	8	18	1	7	2	8	4	5	6	3
9	3	2	1	4	6	5	8	7	19	3	8	2	5	4	7	6	1
10	4	1	7	3	2	5	8	6	20	1	6	2	5	4	3	8	7

Table 13.3
Response frequencies of rankings of products

	Rank							
Product	1	2	3	4	5	6	7	8
A	6	1	6	4	0	1	0	2
B	8	2	0	1	4	1	3	1
C	2	7	0	1	2	3	3	2
D	0	3	4	5	3	4	0	1
E	2	3	2	5	3	5	0	0
F	0	1	4	3	7	1	4	0
G	0	0	0	1	0	4	3	12
H	2	3	4	0	1	1	7	2

From Figure 13.2, three possible groups are indicated: (G), (A, B, C, H) and (D, E, F). Separation of the last two groups is made using the quadratic effects axis, and would not have been possible using the usual Friedman test and multiple comparisons approach described in Conover (1998), which uses only the linear axis. From the M_i and V_i values displayed in Figure 13.2, there is consensus that G ranks poorly (V_i is large and positive, and by inspection of Table 13.3 this indicates clumping at one end, ie consensus, while the large M_i indicates the poor ranking). Average ranks are received by E, D and F (M_i are small). This is in agreement with 'by eye' inspection of the frequency distributions for these products given in Table 13.3. Given the negative corresponding V_i s, and inspection of Table 13.3, there is possibly some evidence of segmentation for A, B, C and H. Remember, as in the previous paragraph, that Figure 13.2 plots $M_i \sqrt{(t-1)/t}$ and $V_i \sqrt{(t-1)/t}$.

The 95% confidence circle indicates product G is significantly different from what would be expected if there were no product differences. Products A, B, D and F are also different from what would be expected if there were no differences, but only just. Further discussion of this data is given, for the statistically minded, in Best and Rayner (1997).

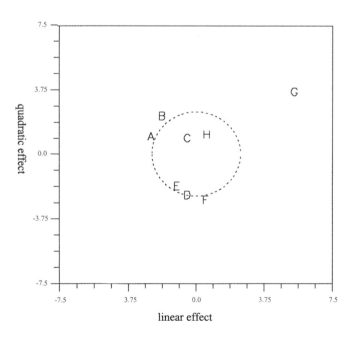

Figure 13.2
Product map and 95% confidence circle for the product ranking data.

For ranking data, $((t-1)/t)\sum_{i=1}^{t} M_i^2$ is precisely the usual Friedman rank test statistic for randomised block data. Recall that for scores 1, 2, ... , k for the category data of Table 13.1, $\sum_{i=1}^{t} M_i^2$ was very similar to the Kruskal-Wallis statistic adjusted for ties. If needed, Windows 95 or Windows NT software to calculate M_i, V_i is available from the first author.

APPENDIX

Let scores $j = 1, ... , k$ be assigned to the categories where, for example, $k = 5$ for the Table 13.1 data. Now define:

$g_1(j) = X\{j-S_1\}$,

$g_2(j) = Z\{j^2-XYg_1(j)-S_2\}$,

N_{ij} = count for the (i, j)th cell of the table of counts, eg Table 13.1

$n = \sum_{i=1}^{t} \sum_{j=1}^{k} N_{ij}$

in which t is the number of products, ie t=5 for Table 13.1

$S_u = n^{-1} \sum_{j=1}^{k} j^u N_{ij}$

$X = (S_2-S_1^2)^{-0.5}$

$Y = S_3-S_1S_2$

$Z = (S_4-S_2^2-X^2Y^2)^{-0.5}$

$M_i = \sum_{j=1}^{k} N_{ij}g_1(j)/\sqrt{n}$ a linear or mean effect, and

$V_i = \sum_{j=1}^{k} N_{ij}g_2(j)/\sqrt{n}$ a quadratic or dispersion effect.

REFERENCES

Baba, Y. (1994) New approaches based on ranking in sensory evaluation. In *New Approaches in Classification and Data Analysis.* Diday, E., Lechevallier, Y., Schader, M., Bertrand, P. and Burtschy, B. (eds), Springer-Verlag, NY.

Best, D.J. and Rayner, J.C.W. (1997) Product maps for ranked preference data. *J.R.S.S., Series D (The Statistician)*, 46(3), 347–354.

Conover, W.J. (1998) *Practical Nonparametric Statistics* (3rd ed). Wiley, NY.

Hirotsu, C. (1990) Discussion of Hamada and Wu. *Technometrics*, 32, 133–136.

HOW SENSORY CELLS ENCODE INFORMATION: THE PROCESSES THAT UNDERLIE SENSITIVITY TO CHEMICAL STIMULI, THEIR QUALITY AND THEIR QUANTITY

P.H. BARRY, S. BALASUBRAMANIAN AND J.W. LYNCH

INTRODUCTION

The olfactory system plays an important role in enabling an appreciation of, and reaction to, the world around us: in our enjoyment of the fragrance of such objects as flowers and perfumes; in reproduction (particularly in animals), and in mother-offspring bonding; in the appreciation of food; in evaluating food that has deteriorated; and in warning us about the presence of noxious substances. That a very large part (about 1%) of the human genome is involved in odour detection is evidence of its significance (Axel, 1995). The sensitivity of the olfactory system is also extremely impressive, exceeding the sensitivity of any other neuronal system: detecting odorants down to molecular, picomolar concentrations (10^{-12} M). Furthermore, the selectivity of the olfactory system is exquisite: distinguishing between at least 10,000 different odorants, with many differently perceived odours only having very minor differences in molecular structure. Most common smells represent complex mixtures of odorants.

ORGANISATION OF THE OLFACTORY SYSTEM

Odorants pass up the nasal passages to a layer of specialist cells known as the olfactory epithelium. They can also pass by a second group of olfactory cells, the vomeronasal organ, that is specially geared to detect pheromones important in reproduction. The olfactory epithelium (Figure 14.1) is covered by a thin layer of mucus in which odorants are dissolved. As the odorants pass along the olfactory epithelium, they can bind to special receptor molecules on hair-like

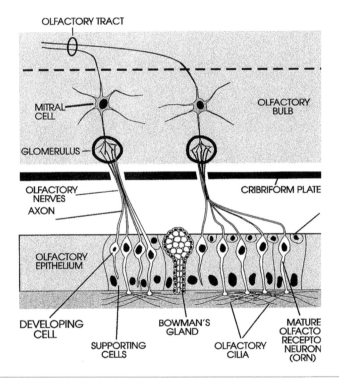

Figure 14.1

Schematic diagram of some of the connections between the olfactory epithelium and olfactory bulb in the olfactory system. The connections between olfactory receptor neurons (ORNs) and glomeruli are not intended to represent any particular organisational links, but merely to indicate that each ORN sends an axon to one glomerulus within the olfactory bulb. In reality, Sullivan and Dryer (1996) have shown that ORNs from different localities, but with a particular expressed receptor type, each project to the same glomerulus (as indicated by Sullivan, in this volume). (Redrawn from a figure of Mervat Hallani, personal communication).

cilia (about 5–20 per neuron) that project out from knobs at the end of a highly specialised neuron, the olfactory receptor neuron (ORN). These ORNs, of which we have about 10^7, have axons that enter the brain to converge on glomeruli in the olfactory bulb. About 10,000 axons converge on each of about 1000 glomeruli present in the olfactory bulb. From here, the signal from the olfactory bulb passes higher up in the brain to the olfactory cortex, where it is finally decoded.

In addition to their specialised structure, these ORNs are also unique amongst neurons (1) in that they are the only neurons to be regenerated throughout the life of an animal and (2) in their extremely high input resistance; electrically about 10 to 100 times higher resistance per unit area than most other neurons.

HOW CAN WE INVESTIGATE THE PRINCIPLES UNDERLYING OLFACTORY RECEPTION?

One of the major technological breakthroughs for investigating the mechanisms of olfactory reception at the cellular level has been the patch-clamp technique (Hamill *et al.*, 1981). This technique allows the direct measurement of the properties of individual ionic channels in intact cells, in excised patches of cell membranes, or measurement of the electrical properties of whole cells (eg, Figure 14.2).

HOW DO THESE CELLS ENCODE AND DECODE OLFACTORY INFORMATION?

Genetic studies of Buck and Axel have suggested that there are about 1000 different odour receptors, with each ORN probably only expressing one type of receptor (Axel, 1995). When an odorant molecule binds to a particular receptor, a signal cascade is initiated, involving one of two second messenger pathways: the adenosine 3',5'-cyclic monophosphate (cAMP), or the inositol 1,4,5,-trisphosphate (IP_3) systems.

THE cAMP SYSTEM

The cAMP system (Figure 14.3 on page 124) is, of the two, the more clearly established pathway experimentally. Odorant binding to the cAMP type receptor causes a membrane-bound G protein (G_{olf}) to bind to, and activate, a membrane-bound adenylate cyclase that catalyses the formation of cAMP from ATP. The cAMP then diffuses and binds to a cyclic nucleotide-gated (CNG) protein channel. The binding of cAMP to the CNG channel opens it, and lets Na^+ and Ca^{2+} ions move down their electrochemical gradients into the cell, thus depolarising it.

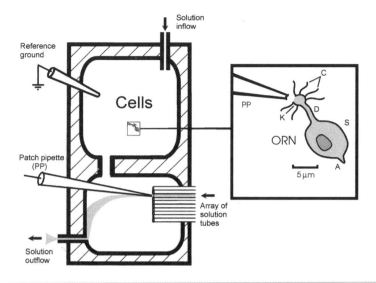

Figure 14.2
One type of bath to explore properties of small excised inside-out patches of membrane from the dendritic knobs of mammalian ORNs and to allow their movement between different test solution flows. In the inset showing an ORN being patch-clamped, C refers to cilia, K to dendritic knob, D to dendrite, S to soma and A to the remains of an axon. (Redrawn from Figure 1A of Balasubramanian et al., 1995).

In the larger, amphibian ORNs, it has been possible to even patch and excise sections of cilial membrane and demonstrate the existence of such CNG channels in that membrane. Although this has not been possible with mammalian ORNs, because of their much smaller size, it has been possible (with difficulty) to patch and excise membranes from the very small dendritic knobs of these ORNs. This has allowed direct measurements of the properties of these CNG channels, in response to the application of cAMP to the inside surface of membrane patches (Figure 14.2). Such measurements have demonstrated that these channels are highly cation selective, and that they have some permeation properties similar to ligand-gated ion channels like the acetylcholine (ACh) receptor channel. These properties include a very low anion permeability, and a minimum cross-sectional area of about 6.5 x 6.5 Å (indicated by their permeability to certain organic cations like Tris and TEA) (Balasubramanian et al., 1995, 1997).

Unlike the ligand-gated receptor channels, CNG channels do not exhibit any desensitisation to their agonist, cAMP. At high external concentrations (many mM), Ca^{2+} ions tend to block these channels,

Figure 14.3
A schematic
diagram of the
cAMP transduc-
tion pathway in
olfactory recep-
tor neurons.
(Partly based on
Figure 9 of
Balasubramanian
et al., 1996).

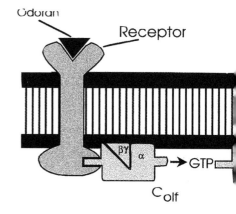

CBP = calcium-binding protein

CAM = calmodulin

Figure 14.4
A schematic dia-
gram of the mol-
ecular structure
of cyclic
nucleotide-gated
(CNG) channels,
to show the
basic structure of
each individual
subunit and a
top view of the
tetrameric com-
bination of sub-
units to form an
ion channel
(inset). (Based
on Figure 1 of
Zagotta and
Siegelbaum,
1996).

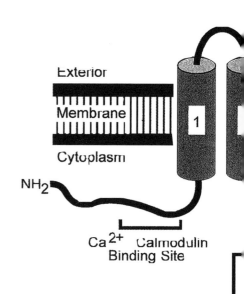

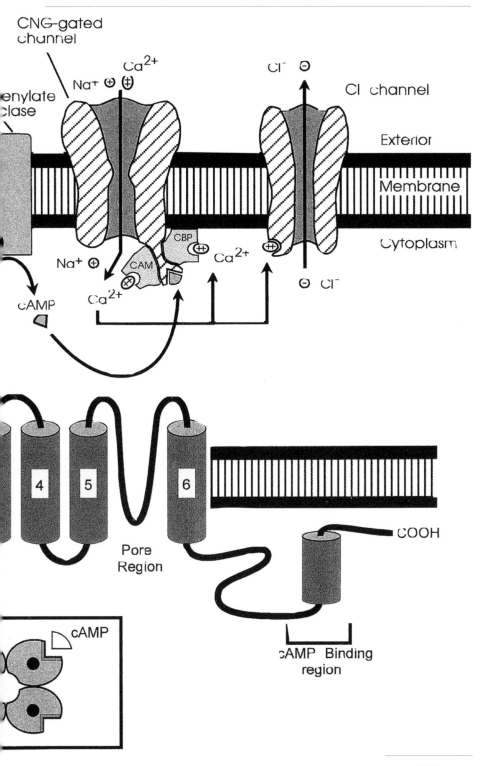

whereas at low concentrations they permeate readily through them (Lynch and Lindemann, 1994). As Ca^{2+} ions enter the cell through these channels:

1   they can bind to and activate some Cl^- channels, which allow Cl^- ions to leave the cell and cause further depolarisation (Kleene and Gesteland, 1991; Lowe and Gold, 1993; Kurahashi and Yau, 1994) and;

2   they can also bind to two proteins: calmodulin (CAM), and a membrane-attached calcium-binding protein (CBP), both of which reduce the cAMP activation of the CNG channels and therefore, following extended odorant activity, help to turn off these channels (Balasubramanian et al., 1996).

For the properties of these Cl^- channels see Larsson et al. (1997) and Hallani et al. (1998).

What else do we know about the CNG channels? They have been cloned and their structure is known (Zufall et al., 1994; Zagotta and Siegelbaum, 1996). They are comprised of six membrane-spanning domains, together with a hydrophilic aqueous pore region. By analogy with K^+ channels, with which they share much homology, these CNG channels probably form tetramers with combinations of both α and β subunits (Figure 14.4). The CAM and cAMP binding sites are both found on the cytoplasmic ends of the protein subunits.

The IP$_3$ system

The precise role of the other second messenger system (IP_3) has not been so unequivocally established, though there is no doubt that IP_3 is elevated in some ORNs following activation by certain odorants (Restrepo et al., 1996). It is suggested that certain odorants, when bound to IP_3 receptors, again activate a membrane-bound G protein to bind to a phospholipase C, and catalyse the formation of IP_3. This, in turn, causes the opening of a Ca^{2+} channel, a non-specific cation channel and a Ca^{2+}-activated K^+ channel. IP_3 may also act via the release of Ca^{2+} from intracellular stores, and, where both it and cAMP second messenger systems are present in a cell, there may be cross-talk between the two systems.

ORN ACTIVATION

On the *activation* side, the entry of Na^+, Ca^{2+} and exit of Cl^- ions all cause initial depolarisation of the ORN. As patch-clamp measurements have indicated, these ORNs are extremely high resistance cells, and under some patch-clamp conditions the opening up of one channel in a membrane patch can initiate an action potential in the rest of the cell (Lynch and Barry, 1989). *In vivo*, these ORNs should also be very responsive to the depolarisation caused by the opening of the above CNG channels. The resultant action potential, probably initiated in the initial segment of the cell where the axon meets the cell soma, is the electrical signal propagated to the olfactory bulb and beyond. Prolonged odour activation should give rise to a train of action potentials.

How is such a signal modulated and terminated? As mentioned above, the entry of Ca^{2+} ions through the CNG channels (and possibly also through some voltage-activated Ca^{2+} channels) via the binding of CAM and CBP, reduces the sensitivity for cAMP activation of the CNG channels. In addition, there are a number of different K^+ channels that have been identified and characterised in membranes of the soma of mammalian ORNs. These include: the delayed rectifier K^+ channel present in most neurons; a transient, rapidly inactivating K^+ channel; a slowly activating K^+ channel; a Ca^{2+} activated K^+ channel; and an inward rectifier K^+/Na^+ channel (Lynch and Barry, 1992). These channels must help to modulate action potential signal spacing, and to terminate the signal completely, thus limiting the detection of odour quantity.

Even though there might be about 1000 different receptors, and it is considered that each ORN only expresses one type of receptor, individual ORNs do respond to some extent to quite a number of different odorants. This, of course, must be the case, since there are in excess of 10,000 different odours that can be detected (Axel, 1995). Receptors are clearly not highly tuned to respond to only single odorants. For example, in some intact patch measurements, a cocktail of five odorants (cineole, n-amyl acetate, methyl salicylate, limonene and α-pinene) caused a response in about 38% of cells (Chiu et al., 1997). There was also a considerable delay between odorant perfusion and cell depolarisation, which in these experiments probably reflected the delay between the generation of the cAMP in the cilia and dendritic knob, and the diffusion of cAMP to the CNG channels in the cell soma.

HOW DO WE DISTINGUISH A PARTICULAR ODOUR QUALITY?

It is interesting that the number of glomeruli are of the same order as the number of distinct receptors (about 10^3). Evidence clearly suggests that although there is some spatial organisation of receptor types within the olfactory epithelium into at least four major regions, within those regions neurons expressing a particular receptor are somewhat randomly located (see Sullivan, 1998; this volume). In addition, what is probably even more significant is that it has also been shown that axons from the same receptor type are connected to only about one or two glomeruli specific for that receptor in the olfactory bulb (Axel, 1995; Sullivan and Dryer, 1996). Thus, the pattern of glomeruli being activated will be characteristic for a particular odorant, and hence the olfactory cortex will decode this pattern in terms of a particular recognised odour (see also Malnic et al., 1999).

ACKNOWLEDGEMENT

We would like to acknowledge the support of the Australian Research Council.

REFERENCES

Axel, R. (1995) The molecular logic of smell. Sci. Am., Oct., 130–137.

Balasubramanian, S., Lynch, J.W. and Barry, P.H. (1995) The permeation of organic cations through cAMP-gated channels in mammalian olfactory receptor neurons. J. Membr. Biol., 146, 177–191.

Balasubramanian, S., Lynch, J.W. and Barry, P.H. (1996) Calcium-dependent modulation of the agonist affinity of the mammalian olfactory cyclic nucleotide-gated channel by calmodulin and a novel endogenous factor. J. Membr. Biol., 152, 13–23.

Balasubramanian, S., Lynch, J.W. and Barry, P.H. (1997) Concentration dependence of sodium permeation and sodium ion interactions in the cyclic AMP-gated channels of mammalian olfactory receptor neurons. J. Membr. Biol., 159, 41–52.

Chiu, P., Lynch, J.W. and Barry, P.H. (1997) Odorant-induced currents in intact patches from rat olfactory receptor neurons: theory and experiment. Biophys. J., 72, 1442–1457.

Hamill, O.P., Marty, A., Neher, E., Sakmann, B. and Sigworth, F.J. (1981) Improved patch-clamp techniques for high-resolution current recording from cells and cell-free membrane patches. Pflügers Arch., 391, 85–100.

Hallani, M., Lynch, J.W. and Barry, P.H. (1998) Characterization of calcium-activated chloride channels in patches excised from the dendritic knob of mammalian olfactory receptor neurons. J. Membr. Biol., 161, 163-171.

Kleene, S.J. and Gesteland, R.C. (1991) Calcium-activated chloride conductance in frog olfactory cilia. J. Neurosci., 11, 3624–3629.

Kurahashi, T. and Yau, K-W (1994) Olfactory transduction: Tale of an unusual chloride current. Curr. Biol., 4, 256–258.

Larsson, H.P., Kleene, S.J. and Lecar, H. (1997) Noise analysis of ion channels in non-space clamped cables: estimates of channel parameters in olfactory cilia. Biophys. J., 72, 1193–1203.

Lowe, G. and Gold, G.H. (1993) Nonlinear amplification by calcium-dependent chloride channels in olfactory receptor cells. Nature, 366, 283–286.

Lynch, J.W. and Barry, P.H. (1989) Action potentials initiated by single channels opening in a small neuron (rat olfactory receptor). Biophys. J., 55, 755–768.

Lynch, J.W. and Barry, P.H. (1992) Studying olfactory transduction using patch-clamping. Today's Life Science, 4, 26–42.

Lynch, J.W. and Lindemann, B. (1994) Cyclic nucleotide-gated channels of the rat olfactory receptor cells: divalent cations control the sensitivity to cAMP. J. Gen. Physiol., 103, 87–106.

Malnic, B., Hirono, J., Sato, T. and Buck, L.B. (1999) Combinational receptor codes for odors. *Cell*, 96, 713–723.

Restrepo, D., Teeter, J. and Schild, D. (1996) Second messenger signalling in olfactory transduction. *J. Neurobiol.*, 30, 37–48.

Sullivan, S.L. and Dryer, L. (1996) Information processing in mammalian olfactory system. *J. Neurobiol.*, 30, 20–36.

Zagotta, W. N. and Siegelbaum, S.A. (1996) Structure and function of cyclic nucleotide-gated channels. *Annu. Rev. Neurosci.*, 19, 235–63.

Zufall, F., Firestein, S. and Shepherd, G.M. (1994) Cyclic nucleotide-gated ion channels and sensory transduction in olfactory receptor neurons. *Annu. Rev. Biophys. Biomol. Struct.*, 23, 577–607.

INFORMATION CODING IN THE MAMMALIAN OLFACTORY SYSTEM

S.L. SULLIVAN

Mammals have an acute olfactory system capable of discriminating among thousands of structurally diverse odorant molecules. To achieve this immense discriminatory capability, the olfactory system employs several molecular and organisational strategies to encode incoming sensory information. In this chapter, I will review these strategies and the processing of olfactory information from the initial step of ligand-receptor protein binding to the transmission of information to the first relay in the brain, the olfactory bulb.

As an odorant enters the nasal cavity, it binds to odorant receptors (ORs) present on the cilia of the olfactory sensory neurons (OSNs), which are the primary sensory neurons of the olfactory system. Binding of odorants to their receptors triggers a cascade of events that ultimately lead to depolarisation of the OSNs, and propagation of action potentials along their axons. The unmyelinated OSN axons gather to form the olfactory nerve, which carries peripheral input to a series of spherical elements in the olfactory bulb, known as glomeruli. Within glomeruli, peripheral axons form synapses with mitral and tufted (M/T) cells, the output cells of the olfactory bulb. Following refinement by local bulbar circuitry, information is then carried via the axons of the M/T cells to higher brain regions where the conscious perception of odorants occurs (for details see Shepherd and Greer, 1997).

ODORANT RECEPTORS

Odorant receptors are G-protein coupled receptors, encoded by a large multigene family (Buck and Axel, 1991). It is estimated that there are as many as 500 to 1000 different OR genes in both mouse and human. Sequence comparisons among members of the OR gene family indicate that although they share some sequence motifs, they are highly divergent. The large size and tremendous diversity of the OR gene family indicate that the olfactory discrimination relies heavily on the different binding properties of individual receptor types.

A recent functional study of the binding properties of an individual OR indicates that the receptive field of an OR is very narrow (Zhao et al., 1998). One receptor, I7, when overexpressed in the olfactory neuronal population, led to an increase in sensitivity to a small subset of aldehydes with carbon chain lengths between seven and ten. Aldehydes with longer or shorter chain lengths failed to elicit responses. These results are in accord with studies of the response profiles of individual mouse OSNs, each of which is thought to express a single OR gene. Using varying concentrations of two series of related compounds, Sato et al. (1994) demonstrated that the specificities of OSNs are very narrow and sharpen with decreasing concentrations of odorants. At low concentrations (of 1 µM), OSNs were found to be highly selective for particular molecular features of odorants, including length of hydrophobic domains, electrical charges, and electronegativities of atoms.

These studies suggest that the features of an odorant are extracted and encoded by the types of ORs with which it interacts, and the activation of OSNs expressing these receptors. In other words, the olfactory system may 'view' an odorant as a composite of its unique set of molecular determinants, and the unique set of ORs that bind these determinants. Such thinking is in contrast to earlier notions that OSNs are broadly tuned and respond to a wide range of odorants. The basis for the apparent broad specificity of OSNs to different odorants may be explained in part by the high concentrations of odorants used in some studies, and in part by ORs being highly specific for molecular features yet binding many odorants that have these features in common.

EXPRESSION PATTERNS OF ODORANT RECEPTORS IN THE OLFACTORY EPITHELIUM

How does the olfactory system organise the information extracted from hundreds of different receptor types? A number of studies have examined the extent to which input received by a specific receptor type is spatially organised, at the levels of both the olfactory epithelium and the olfactory bulb.

In situ hybridisation studies examining the spatial expression patterns of ORs in the olfactory epithelium, indicate that each OR gene is expressed by a small subset of OSNs (Ressler *et al.*, 1993; Vassar *et al.*, 1993; Strotmann *et al.*, 1994). On average, each OR gene is expressed in 0.1% to 0.2% of the total olfactory sensory neuronal population, which corresponds to approximately 5000–10,000 OSNs. In examining the positions of neurons expressing a given OR, it was noted that these neurons are generally confined to one of four expression zones within the olfactory epithelium (Figure 15.1). Each expression zone occupies a different domain along the dorsal-ventral, medial-lateral axes of the nasal cavity and encompasses approximately 25% of the olfactory epithelial surface area.

Within an expression zone, there is no further organisation of OSNs expressing a given OR. OSNs expressing a given OR are not clustered in the epithelium, but are broadly distributed and surrounded by many OSNs expressing different receptor types. Therefore, although the features of an odorant are precisely encoded by the molecular specificities of the receptors, there is little spatial organisation of this information within the olfactory epithelium. Rather, the information is only broadly organised into four zonal sets. Although the function of the segregation of information into zones is unclear, it is maintained in the olfactory bulb with each zone of the epithelium projecting to a different dorsal-ventral domain of the olfactory bulb. However, unlike the zones of the olfactory epithelium, information is highly organised within the domains of the olfactory bulb.

ACCESSORY OLFACTORY SYSTEM

Mammals have a second anatomically distinct chemosensory system, the accessory olfactory system, which processes information

A. Spatial representation of information in the olfactory epithelium and bulb

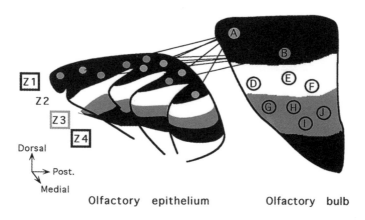

Figure 15.1
Schematic representation of olfactory information coding. Incoming olfactory information is broadly organised into four spatial zones (Z1–Z4) within the olfactory epithelium. In a lateral view of the olfactory epithelium, zones appear as bands organised along the dorsal-ventral axis. Within an expression zone, neurons expressing the same OR gene (indicated by like colors) are widely distributed. The zonal organisation of incoming information is preserved in the olfactory bulb with each expression zone projecting to a different broad domain of the olfactory bulb. Within a domain of the bulb, however, information is highly refined. Despite their wide distribution in the olfactory epithelium, neurons expressing the same OR gene project to one medial and one lateral glomerular site (A-L) in each olfactory bulb. For simplicity, only the medial aspect of the bulb is schematised.

in parallel to the main olfactory system and is thought to be receptive to both volatile and non-volatile ligands (for review see Wysocki and Meredith, 1987). The primary sensory neurons of the accessory olfactory system are contained within the vomeronasal organ, which is a tubular structure located on the ventral portion of the nasal septum. These neurons project their axons to the accessory olfactory bulb located on the dorsal-posterior aspect of the olfactory bulb.

Recent studies indicate that there are several similarities between the main and the accessory olfactory systems with respect to information processing. Two putative chemosensory receptor families have been identified; each consisting of 100–200 members which are highly expressed in the sensory neurons of the vomeronasal organ (Dulac and Axel, 1995; Herrada and Dulac, 1997; Matsunami and Buck, 1997; Ryba and Tirindelli, 1997). Like the ORs of the main olfactory system, both of these gene families encode G-protein coupled receptors. Members of these families, however, are highly divergent from one another, as well as from the ORs of the main olfactory system. Analogous to the expression zones of the olfactory epithelium, the two identified families of vomeronasal receptors are expressed in a distributed manner in distinct zones of the vomeronasal organ. These zones are characterised not only by the expression of different families of receptors, but also by different molecular markers including G protein a subunits (Halpern *et al.*, 1995; Berghard *et al.*, 1996). The zonal projection pattern observed between the olfactory epithelium and the olfactory bulb is also observed in the accessory olfactory system. The axonal projections of primary neurons in the two different vomeronasal zones remain segregated in the accessory olfactory bulb (Jia and Halpern, 1996). The extent to which information is further refined spatially within the accessory olfactory bulb has not yet been determined.

GLOMERULAR MAP OF OLFACTORY INFORMATION IN THE MAIN OLFACTORY BULB

OSNs extend axons from the olfactory epithelium to the olfactory bulb. As these axons reach the olfactory bulb they travel in the olfactory nerve layer of the bulb, until they enter the bulb and make synaptic contacts with secondary neurons in anatomically distinct structural units called glomeruli (for review see Shepherd and Greer, 1997). There is a considerable amount of anatomical convergence at the level of the olfactory epithelium to bulb projection. In the mouse, there are approximately 2000 glomeruli, each of which receives input from one to two thousand olfactory neurons and shares information with 25–50 M/T cells. Within the main olfactory system, each mitral cell and each OSN innervates a single glomerulus.

Several studies using a variety of methods — including the uptake of 2-deoxyglucose, induction of c-fos, and the activation of voltage sensitive dyes — have examined the glomerular activation patterns in the olfactory bulb induced by exposure to different odorants (for review see Shepherd, 1994). The major conclusions drawn from these studies are that individual glomeruli generally respond to more than one odorant, and that a given odorant generally activates more than one glomerulus. However, each odorant elicited a unique pattern of glomerular activity in the bulb suggesting that glomeruli serve as functional units or coding modules in the processing of olfactory information.

Recently, the molecular basis for glomeruli serving such a role has been presented (Ressler *et al.*, 1994; Vassar *et al.*, 1994; Mombaerts *et al.*, 1997). Two approaches, *in situ* hybridisation analysis of OR mRNA present in the axonal terminals of OSNs, and gene replacement experiments in which the expression of lacZ and specific OR genes are linked, have been used to map the projection patterns of OSNs expressing a given receptor type. These studies demonstrated that the axons of OSNs expressing the same OR, converge onto one of two glomerular sites in each olfactory bulb. Invariably, one of these sites was located on the medial aspect of the bulb and the other on the lateral aspect. This convergence of axons of neurons expressing the same OR results in a precise spatial map of olfactory information within the glomerular layer of the bulb, with information amassed by thousands of OSNs in the olfactory epithelium focused onto two glomerular sites in each bulb. The glomerular spatial map is bilaterally symmetric between the two olfactory bulbs, and is conserved among different animals.

Together with the functional data, these studies indicate that at the level of the olfactory bulb, the response to an odorant is stereotyped and represented by the activation of a unique set of glomeruli (Figure 15.2; see also Kauer, 1987; Mori and Shepherd, 1994). In simplest terms, the extraction of odorant features, which begins with the differential binding properties of ORs, is directly reflected spatially in the olfactory bulb through the existence of anatomically distinct glomeruli, and associated mitral cells, dedicated to receiving information from a single receptor type. Studies of the electrophysiological response of individual mitral cells are consistent with this model, and indicate that mitral cells are highly selective for particular molecular determinants and respond to structurally similar odorants that have these determinants in common (Mori and Yoshihara, 1995). As opposed to an individual glomerulus being specific for a single odorant, the use of combinations of glomeruli to encode the features of an odorant allows for a much greater number of odorants to be discriminated. Furthermore, the discriminatory capability of the olfactory system would greatly exceed genomic constraints placed on the size of the OR gene family as well as anatomical constraints placed on the number of glomeruli and, consequently, the size of the olfactory bulb.

The field of olfaction has come far in the last few years, and we now have a better understanding of the molecular and cellular bases of olfactory discrimination. Future advances in our knowledge of

B. Glomerular map of olfactory information

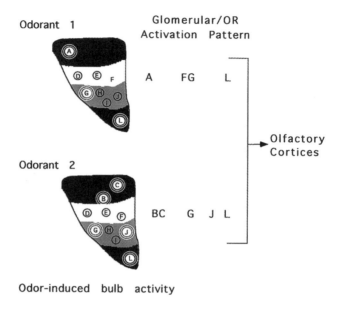

Odorant 1

Glomerular/OR
Activation Pattern

A   FG   L

Olfactory
Cortices

Odorant 2

BC   G   J   L

Odor-induced bulb activity

Figure 15.2
Glomerular map of olfactory information. At the level of the olfactory bulb, the response to an odorant consists of the activation of glomeruli (indicated in white) receiving input from receptors that bind the odorant. Since a given odorant may display multiple molecular determinants recognised by different receptors, the bulb response may consist of the activation of multiple glomeruli. Furthermore, different odorants with molecular determinants in common may activate some of the same glomeruli (eg, G and L). However, the overall glomerular activation pattern for each odorant is expected to be unique and form the basic output from the bulb to olfactory cortices.

olfactory perception lie in the anatomical tracing of the axonal projections from the olfactory bulb to the cortex, where the conscious perception of olfaction occurs, and the imaging of cortical responses to odour stimulation.

REFERENCES

Berghard, A. and Buck, L.B. (1996) Sensory transduction in vomeronasal neurons: Evidence for G alpha o, G alpha i2, and adenylyl cyclase II as major components of a pheromone signalling cascade. *J. Neurosci.*, 16, 909–918.

Buck, L. and Axel, R. (1991) A novel multigene family may encode odorant receptors: a molecular basis for odor recognition. *Cell*, 65, 175–187.

Dulac, C. and Axel, R. (1995) A novel family of genes encoding putative pheromone receptors in mammals. *Cell*, 83, 195–206.

Halpern M., Shapiro, L.S., and Jia, C.P. (1995) Differential localization of G-proteins in the opossum vomeronasal system. *Brain Res.,* 677, 157–161.

Herrada, G. and Dulac, C. (1997) A novel family of putative pheromone receptors in mammals with a topographically organised and sexually dimorphic distribution. *Cell,* 90, 763–773.

Jia, C.P., and Halpern, M. (1996) Subclasses of vomeronasal receptor neurons: Differential expression of G proteins (G(i alpha 2) and G(o alpha)) and segregated projections to the accessory olfactory bulb. *Brain Res.,* 719, 117–128.

Kauer, J.S. (1987) Coding in the olfactory system. In Finger, T.E. and Silver, W.L. (eds), *The Neurobiology of Taste and Smell,* Wiley, NY, pp. 205–231.

Matsunami, H. and Buck, L. (1997) A multigene family encoding a diverse array of putative pheromone receptors in mammals. *Cell,* 90, 775–784.

Mombaerts, P., Wang, F., Dulac, C., Chao, S.K., Nemes, A., Mendelsohn, M., Edmondson, J., and Axel, R. (1997) Visualizing an olfactory sensory map. *Cell,* 87, 675–686.

Mori, K. and Shepherd, G. M.(1994) Emerging principles of molecular signal processing by mitral/tufted cells in the olfactory bulb. *Semin. Cell Biol.,* 5, 65–74.

Mori, K. and Yoshihara, Y. (1995) Molecular recognition and olfactory processing in the mammalian olfactory system. *Prog. Neurobiol.,* 45, 585–619.

Ressler, K.J., Sullivan, S.L., and Buck, L.B. (1993) A zonal organization of odorant receptor gene expression in the olfactory epithelium. *Cell,* 73, 597–609.

Ressler, K.J., Sullivan, S.L., and Buck, L.B. (1994) Information coding in the olfactory system: Evidence for a stereotyped and highly organised epitope map in the olfactory bulb. *Cell,* 79, 1245–1255.

Ryba, N.J.P., and Tirindelli, R. (1997) A new multigene family of putative pheromone receptors. *Neuron,* 19, 371–379.

Sato, T., Hirono, J., Tonoike, M., and Takebayashi, M. (1994) Tuning specificities to aliphatic odorants in mouse olfactory neurons and their local distributions. *J. Neurophysiol.,* 72, 2980–2989.

Shepherd, G. M. (1994) Discrimination of molecular signals by the olfactory receptor neurons. *Neuron,* 13, 771–790.

Shepherd, G. M. and Greer, C. A. (1997) Olfactory bulb. In Shepherd, G.M. (ed), *The Synaptic Organization of the Brain.* Oxford University Press, NY, pp. 159–203.

Strotmann, J., Wanner, I., Helfrich, T., Beck, A., and Breer, H. (1994) Rostro-caudal patterning of receptor expressing neurones in the rat nasal cavity. *Cell Tiss. Res.,* 278, 11–20.

Vassar, R. , Chao, S. K., Sitcheran, R., Nunez, J. M., Vosshall, L. B., and Axel, R. (1994) Topographic organization of sensory projections to the olfactory bulb. *Cell,* 79, 981–991.

Vassar, R., Ngai, J., and Axel, R. (1993) Spatial segregation of odorant receptor expression in the mammalian olfactory epithelium. *Cell,* 74, 309–318.

Wysocki, C.J. and Meredith, M. (1987) The vomeronasal system. In Finger, T.E. and Silver, W.L. (eds), *The Neurobiology of Taste and Smell.* Wiley, NY, pp. 125–150.

Zhao, H., Lidija, I., Otaki, J. M., Hashimoto, M., Mikoshiba, K., and Firestein, S. (1998) Functional expression of a mammalian odorant receptor. *Science,* 279, 237–242.

ANATOMY OF THE PERIPHERAL CHEMOSENSORY SYSTEMS: HOW THEY GROW AND AGE IN HUMANS

B. KEY

INTRODUCTION

Understanding the anatomical basis of chemosensory function and its behavioural consequences is essential for effective marketing in the food and perfumery industry. There is no value in targeting products to specific age groups if these populations are incapable of responding to specific sensory stimuli. At least three sensory systems play an important role in the appreciation of food and beverages — the olfactory, gustatory and trigeminal systems. In this paper I describe the anatomy of these sensory systems, and highlight how structural changes with age may influence function.

The effect of age on the function of the chemosensory systems associated with perception of flavour is considerable. The ability to detect odours decreases with age, and this is most notable over the age of 70 years (Doty *et al.*, 1984; Wysocki and Gilbert, 1989; Wysocki and Pelchat, 1993). This is consistent with data that indicates that most people find that the perceived intensity of odours declines with age (Wysocki and Pelchat, 1993). In addition, the threshold for odorant detection (or the concentration at which a particular odour can be smelled) is usually higher in the aged (Wysocki and Pelchat, 1993). Age-related deficits in the chemosenses should be of particular concern to the food industry, especially in a country where the population is rapidly aging. Since the flavour of foods is highly dependent on odour, one would predict that eating becomes less appealing with age. However, normal, healthy adults report neither a loss in appetite nor that eating becomes less appealing with age (Stevens, 1989). Instead it would seem that the elderly change their eating habits to accommodate any age-related deficit in their chemosensory systems, as is observed in patients with chemosensory dysfunction (Mattes *et al.*, 1990). However, the effect of chemosensory loss on health and well-being is clearly heterogeneous, with some people suffering severely, as highlighted in the following extract from a personal communication:

> Now I am 74 years of age. I started off about 1 year ago saying all the food tastes like nothing. Maybe it was just the supermarket food. Then I could not smell Pine O Clean or any strong perfumes. Cooking — I burnt everything, it had no smell and no taste. Eating is a big bore, I am never hungry and tend not to eat. I feel, what is the point?

WHAT IS COMMON ABOUT THE ANATOMY OF THE SENSORY SYSTEMS?

In each of the olfactory, gustatory and trigeminal sensory systems, there is a group of specialised cells or neurons that are capable of responding to external stimuli. The olfactory and taste sensory cells are directly exposed to the external environment where they can interact with odours in the air (odorants) and molecules in solution (tastants), respectively. In the trigeminal system, the primary sensory axons terminate freely in the mucosa lining the oral and nasal cavities, and are directly involved in transducing signals regarding the texture, astringency, temperature and pungency of foods. In the gustatory system, the sensory receptor cells are connected to peripheral neurons that, in turn, communicate directly with the central nervous system. In contrast, the receptor cells in the olfactory system are specialised nerve cells that connect directly with the central nervous system.

THE OLFACTORY SYSTEM

In humans, the primary sensory olfactory neurons responsible for detecting odours reside in the olfactory neuroepithelium, which lines the roof of the nasal cavity, the sides of the dorsocaudal nasal septum and the superior turbinate. The remaining nasal cavity is covered by respiratory epithelium. The olfactory neuroepithelium occupies a surface area of about 1 cm^2 on each side of the nose, and has about 30 000 primary sensory neurons per square millimetre, which equates to about 6 million olfactory neurons in total.

Using a special biopsy needle, small pieces of respiratory epithelium and olfactory neuroepithelium can be extracted from the living nasal cavity and processed for histological examination (Moran *et al.*, 1982). The respiratory epithelium is pseudeostratified and consists of a basal layer of stem cells, and an overlying layer of ciliated cells interspersed with goblet cells (Moran *et al.*, 1982). The goblet cells secrete mucus on the surface of the epithelium and this mucus bathes the motile cilia of the ciliated cells. The olfactory neuroepithelium is also pseudostratified, and consists of a basal layer of proliferating stem cells, a middle layer of primary sensory olfactory neurons, and an apical layer of supporting (sustentacular) cells.

The neuroepithelium is separated from the underlying lamina propria by a basal lamina through which the primary olfactory axons pass en route to the brain. Scattered throughout the neuroepithelium are large Bowman's glands that secrete mucus onto the surface of the tissue. The primary olfactory neurons are bipolar and possess an apical dendrite that ends at the surface of the epithelium in a bulbous knob containing 10–30 long, and non-motile, cilia (Moran *et al.*, 1982). These cilia form a dense mat and are embedded in a mucous layer. Odorants in the nasal cavity interact with the receptor proteins present in cilia by first diffusing into the mucous layer, which provides the appropriate ionic environment for odorant-receptor binding.

The axons of primary sensory olfactory neurons exit the basal lamina either individually, or in small fascicles (Moran *et al.*, 1982; Morrison and Costanzo, 1990). The intraepithelial axon fascicles are ensheathed by processes of supporting cells and some basal cells (Morrison and Costanzo, 1992). These axons coalesce into larger bundles in the lamina propria, and project into the cranial cavity through foramina in the cribriform plate of the ethmoid bone that lines the roof of the nasal cavity. The olfactory axons terminate in a specialised protrusion from the ventral brain called the olfactory bulb. Here the axons converge and form glomeruli, tufts of synaptic connections between the terminal arbors of primary sensory olfactory axons, and the dendrites of second-order olfactory neurons. All axons expressing the same type of receptor protein converge onto two glomeruli in each olfactory bulb (Mombaerts et al., 1996). Thus, when the olfactory neuroepithelium is exposed to an odour, specific glomeruli become activated, and it is this combination of glomeruli that forms the basis of the neural code responsible for olfactory discrimination.

Unlike other regions of the nervous system, where neurons are long-lived, in the olfactory system the sensory neurons have a limited life-span. It is generally believed that primary olfactory sensory neurons live for about 30 days, die, and are replaced by proliferating stem cells located in the basal portion of the sensory epithelium. This has led to the idea that these sensory neurons are genetically programmed to die. However, many primary olfactory sensory neurons are long-lived, especially when animals are raised in a clean air environment (Hinds et al., 1984). When the life-span of primary sensory olfactory neurons were more carefully examined in mice, it was revealed that many live for as long as 90 days (Mackay-Sim and Kittel, 1991). It is more likely that these neurons are predisposed to death following environmental insults, rather than possessing a genetically regulated short life-span. Although primary sensory olfactory neurons appear to be quite inaccessible to direct physical perturbations, they can be damaged by head injuries, as well as by volatile toxins, noxious chemicals, and infectious agents. This predisposition to cell death may have evolved as a means of quickly rejuvenating the neuroepithelium so that there is minimal disruption to physiological function following injury.

When primary sensory olfactory neurons die they are replaced by de novo differentiation of neural stem cells that reside in the basal layer of the olfactory neuroepithelium. In normal olfactory neuroepithelia, these basal cells are responsible for most of the proliferative activity (Huard and Schwob, 1995). The stem cells divide, and give rise to a population of proliferating neuronal precursors (or transient amplifying cells). Many die rather than differentiate into neurons. However, there is a surge of proliferative activity in these neuronal precursors following the death of primary olfactory neurons induced by removal of the olfactory bulb (Gordon et al., 1995). These newly generated cells either mature into differentiated neurons, or die if there is no available neural space in the epithelium (Carr and Farbman, 1993; Schwob et al., 1992). About 60% of these transient amplifying cells die between five days and two weeks after their generation (Schwob et al., 1992). There is thus a continual turnover of precursor cells in the neuroepithelium, and these cells only reach maturity to replace dying primary sensory olfactory neurons.

There is particular interest in identifying the signals that mediate neuronal proliferation, differentiation and death in the olfactory neuroepithelium. A better understanding of the underlying mechanisms may lead to an effective therapeutic treatment of olfactory dysfunction in disease and normal aging.

In comparison to other mammals, the olfactory system of humans develops relatively early (Chuah and Zheng, 1992). The olfactory sensory neurons are morphologically mature by the end of the first trimester. By 17 weeks gestation, the first synaptic connections have formed between the olfactory axons and their post-synaptic partners in the olfactory bulb. Glomeruli are morphologically discrete formations at week 23. Physiological testing of olfactory perception in preterm babies indicates that the olfactory system is functional at 28 weeks gestation (Schaal and Orgeur, 1992). However, very little is known about the anatomical complexity of the olfactory nerve pathway at this age. For instance, if we knew how many mature primary sensory olfactory neurons and glomeruli were present at this age, it might provide clues as to the minimal neural composition of the peripheral olfactory system sufficient for olfaction.

The size of the normal aged human olfactory epithelial sheet is considerably reduced in comparison to young adults or children (Nakashima et al., 1984). The most noticeable gross anatomical defect in the aged olfactory neuroepithelium appears to be the presence of numerous patches of non-sensory epithelium interspersed amongst the sensory epithelium (Nakashima et al., 1984; Morrison and Costanzo, 1990; Trojanowski et al., 1991). Thus with age there is a gradual reduction in the surface area of the olfactory neuroepithelium as it is replaced by respiratory epithelium. The mechanisms underlying this loss of olfactory tissue are not known. It could be that the basal stem cells lose their proliferative ability, or that neuronal precursor cells do not differentiate in the aged. On the other hand, it is possible that the rate of primary olfactory neuron death increases in old age to such a level that neural stem cells cannot generate replacements quickly enough.

At a histological level, the cellular arrangement of the aged olfactory neuroepithelium is often highly disorganised (Nakashima et al., 1984). The layering of cell bodies into discrete lamina, that is typical of young epithelia, becomes highly distorted in the elderly. There are also many tangles of nerve fibres distributed in patches throughout the basal regions of the neuroepithelium and submucosa (Trojanowski et al., 1991). These tangles appear to arise from axons that have failed to innervate the olfactory bulb. Interestingly, tangles are also observed in young adults but never in foetal or neonatal olfactory neuroepithelia. The dysfunctional morphology of the olfactory neuroepithelium probably emerges gradually throughout post-natal life.

Animal studies suggest that primary sensory olfactory neurons have an increased incidence of cell death if they fail to form synaptic connections in the olfactory bulb (Schwob et al., 1992). Therefore the formation of tangles of axons from neurons arising from a discrete patch of neuroepithelium may explain why neurons die in patches. Alternatively, it is possible that toxic or infectious agents cause localised trauma in the olfactory neuroepithelium that disrupts regeneration of primary sensory olfactory axons in the elderly. Whatever the underlying mechanism, it seems that these structural changes in the olfactory neuroepithelium are responsible for the age-related deficits in olfactory function. This is supported by evidence that similar morphological alterations are observed in young adults suffering from hyposmia (reduced olfactory sensitivity) (Moran et al., 1992).

THE GUSTATORY SYSTEM

Most regions of the oral cavity are lined by a mucosa consisting of a layer of either keratinised or non-keratinised epithelium overlying a lamina propria (Strachan, 1994). The mucosa of the cheeks, lips and some regions of the palate also contain a submucosal layer of loose connective tissue. The epithelium of the tongue contains numerous folds, ridges and mounds referred to as papillae. There are four types of papillae: filiform, fungiform, vallate and foliate. The filiform papillae, and fungiform papillae (~ 200), are present on the anterior two-thirds of the tongue; the vallate papillae (~ 8–12) form a v-shaped wedge between the anterior two-thirds and the posterior third of the tongue; and the folliate papillae (a series of 2–9 bilateral ridges) are on the posteriolateral surface of the tongue.

Scattered across the epithelium of the tongue, soft palate and epiglottis are taste buds (2000–5000), which are spherical clusters of specialised epithelial cells (70 mm in height and 40 mm in diameter). On the tongue, taste buds are localised to only the fungiform (~ 4 buds/papilla), vallate (~ 250 buds/papilla) and foliate (~ 600 buds/papilla) papillae. Taste buds contain about 40 specialised sensory epithelial cells that possess the receptor proteins and ion channels that directly interact with tastants.

Unlike the olfactory system, where there are hundreds of distinct olfactory receptor proteins and hence functionally distinct types of olfactory sensory cells, there are probably only five different functional types of taste sensory cells. These cells respond to one of the following broad classifications of tastants: salt, sugar, bitter, acid, or umami compounds. Umami (or 'delicious taste') is recognised as a separate primary taste sensation that is elicited by monosodium glutamate (Yamaguchi, 1991). Each taste bud probably contains a mixture of different taste cells, although one particular type is often predominant. Taste cells have a limited life-span of about 10 days, and are continually replaced by proliferating cells in the basal regions of the taste bud. Because of this continual turnover of cells, it is not known whether a taste bud maintains a similar population of taste cells, and hence functional specificity, throughout life. Many texts describe a spatial map of taste buds with different sensitivities across the surface of the tongue. For instance, it is often claimed that the front of the tongue is more responsive to NaCl. However, in general, all regions appear to be equally sensitive to various tastes (Bartoshuk, 1989). The only noticeable difference is that bitterness is often perceived to be more intense in the back of the tongue in the region of foliate papillae (Bartoshuk *et al.*, 1987).

The taste sensory cells are not neurons, as in the olfactory system, but are instead specialised epithelial cells that are innervated by primary sensory axons from three peripheral nerves. The taste buds in the tongue are innervated by either the facial or glossopharyngeal nerves, while the taste buds in the soft palate and epiglottis are innervated by the facial and vagus nerves, respectively. Because of this multiple innervation, complete loss of taste sensation is very rare. Interestingly, even if small areas of oral mucosa lose taste sensation, there is no gross effect on the 'whole mouth' experience of taste (Bartoshuk, 1989).

The first taste buds begin to form *in utero* at about eight weeks in human foetuses, and are mature before birth (Bradley and Stern, 1967). Animal studies indicate that the taste system is functional *in utero*, and that the response characteristics of taste cells changes during development as the system matures and cells gain adult-like phenotypes (Fujimoto *et al.*, 1993; Mbiene and Farman, 1993). Although untested, it is possible that the relative proportion of different subpopulations of taste cells changes throughout life. In general, responsiveness of the taste system in localised regions of the mouth declines with age, but 'whole mouth' sensations are very similar across ages, with the exception of bitterness (Bartoshuk *et al.*, 1986; Bartoshuk, 1989). The reason for this loss is controversial, since it is not clear whether taste bud numbers on tongue papillae change significantly with age. One of the reasons for this lack of clarity seems to be due to the wide variation in taste bud numbers between individuals (Miller and Reedy, 1990). In fact, individuals exhibiting high sensitivity to various tastants have a higher density of taste buds; these people have been referred to as 'supertasters' (Bartoshuk, 1993).

THE TRIGEMINAL SENSORY SYSTEM

The flavour and pleasantness of food is strongly influenced by its temperature, texture, and shape — collectively referred to as 'mouth-feel' (Rolls et al., 1982). These attributes are principally mediated by the trigeminal sensory system (Capra, 1995). The food industry places considerable emphasis on the sensory characteristics of food, and recognises its importance in the overall palatability of its products (Azana et al., 1996; Full et al., 1996). The oral cavity is innervated by branches of the trigeminal, or fifth cranial nerve. The lingual branch of this nerve innervates the tongue and the floor of the mouth, while the nasopalatine and posterior palatine nerves from the mandibular branch of the trigeminal nerve innervate the hard and soft palate, respectively. The buccal nerve, from the mandibular branch of the trigeminal nerve, innervates the cheek mucosa. In the mucosa, primary sensory nerve fibres from the trigeminal nerve either terminate freely or terminate in several types of specialised nerve endings in the mucosa. These fibres either respond directly to chemicals that diffuse into the epithelium, or are stimulated by touch and changes in temperature.

Trigeminal nerve fibres also innervate the olfactory neuroepithelium, where they are stimulated by some inhaled vapours. These fibres are present in both the submucosa, and in the olfactory neuroepithelium, where they course between cells. The trigeminal fibres never reach the apical surface of the neuroepithelium, and are protected from direct access to the external environment in the nasal cavity by tight junctions between the apical margins of epithelial cells (Finger et al., 1990). This explains why membrane-permeable odorants are the most effective trigeminal stimuli in the nose (Silver et al., 1986). In the nasal cavity, the trigeminal nerve is responsible for detecting the pungency of odorants. Interestingly, there is a strong interaction between the trigeminal and olfactory systems, since pungency can diminish the perception of odours (Cain and Murphy, 1980).

Little attention has been given to the development and aging of the trigeminal innervation in the epithelium of either the oral or nasal cavity. Yet this is a very important modality involved in flavour perception (Calhoun *et al.*, 1992). Moreover, sensory discrimination by trigeminal nerves in the mouth is critical for the therapeutic rehabilitation of patients undergoing restorative surgery or radiation therapy. Data is now emerging on the surface sensibilities of the mouth and tongue in both health and disease (Aviv *et al.*, 1992). Oral sensation and perception are also important for the development of normal feeding and swallowing during infancy, and yet we know very little about how food presentation, and types of food, affect learning of feeding skills (Stevenson and Allaire, 1991).

CONCLUSION

Extrapolating from American data (Wysocki and Gilbert, 1989; Smith and Duncan, 1992), it is estimated that at least 0.075% of comparable populations (~ 15 000 Australians) have some form of permanent chemosensory deficit. In order to provide effective therapeutic treatments and/or enhance the sensory experience of foods, we need a comprehensive understanding of the anatomy of the sensory systems in the elderly and sensory-compromised patients. At the other end of the spectrum, we need a good understanding of when these systems become structurally and functionally mature. It may be that limited sensation is achieved before structural maturity, in which case we have an idea of the necessary prerequisites for functional recovery in disorders in adults, or the limitation of the system in the elderly.

REFERENCES

Aviv, J.E., Hecht, C., Weinberg, H., Dalton, J.F. and Urken, M.L. (1992) Surface sensibility of the floor of the mouth and tongue in healthy controls and in radiated patients. *Otolaryngol.*, 107, 418–423.

Azanza, F., Klein, B.P. and Juvik, J.A. (1996) Sensory characteristics of sweet corn lines differing in physical and chemical composition. *J. Food Sci.*, 61, 253–257.

Bartoshuk, L.M. (1989) Taste. Robust against the age span? *Ann. NY Acad. Sci.*, 561, 65–75.

Bartoshuk, L.M. (1993) Genetic and pathological taste variation: what can we learn from animal models and human disease? In Chadwick, D., Marsh, J. and Goode, J. (eds), *The Molecular Basis of Smell and Taste Transduction*. J Wiley, NY, pp. 251–267.

Bartoshuk, L., Desnoyers, S., Hudson, C., Marks, L., O'Brien, M., Catalanotto, F., Gent, J., Williams, D. and Ostrum, K.M. (1987) Tasting on localised areas. *Ann. NY Acad. Sci.*, 510, 166–168.

Bartoshuk, L.M., Rifkin, B., Marks, L.E., Bars, P. (1986) Taste and aging. *J. Gerontol.*, 41, 51–57.

Bradley, R.M. and Stern, I.B. (1967) The development of the human taste bud during the foetal period. *J. Anat.*, 101, 743–752.

Cain, W.S. and Murphy, C.L. (1980) Interaction between chemoreceptive modalities of odour and irritation. *Nature*, 284, 255–257.

Calhoun, K.H., Gibson, B., Hartley, L., Minton, J. and Hokanson, J.A. (1992) Age–related changes in oral sensation. *Laryngoscope*, 102, 109–116.

Capra, N.F. (1995) Mechanisms of oral sensation. *Dysphagia*, 10, 235–247.

Carr, V. McM. and Farbman, A.I. (1993) The dynamics of cell death in the olfactory epithelium. *Exp. Neurol.*, 124, 308–314.

Chuah, M.I. and Zheng, D.R. (1992) The human primary olfactory pathway: fine structural and cytochemical aspects during development and in adults. *Microsc. Res. Tech.*, 23, 76–85.

Doty, R.L., Shaman, P., Applebaum, S.L., Giberson, R., Siksorski, L. and Rosenberg, L. (1984) Smell identification ability: changes with age. *Science*, 226, 1441–1442.

Finger, T.E., St. Jeor, V.L., Kinnamon, J.C. and Silver, W. (1990) Ultrastructure of substance P- and CGRP-immunoreactive nerve fibres in the nasal epithelium of rodents. *J. Comp. Neurol.*, 294, 293–305.

Fujimoto, S., Yamamoto, K., Yoshizuka, M. and Yokoyama, M. (1993) Pre- and post-natal development of rabbit foliate papillae with special reference to foliate gutter formation and taste bud and serous gland differentiation. *Micro. Res. Tech.*, 26, 120–132.

Full, N.A., Reddy, S.Y., Dimick, P.S. and Ziegler, G.R. (1996) Physical and sensory properties of milk chocolate formulated with anhydrous milk fat fractions. *J. Food Sci.*, 61, 1068–1073.

Gordon, M.K., Mumm, J.S., Davis, R.A., Holcomb, J.D. and Calof, A.L. (1995) Dynamics of MASH1 expression in vitro and in vivo suggest a non-stem cell site of MASH1 action in the olfactory receptor neuron lineage. *Mol. Cell. Neurosci.*, 6, 363–379.

Hinds, J.W., Hinds, P.L. and McNelly, N.A. (1984) An autoradiographic study of the mouse olfactory epithelium: evidence for long-lived receptors. *Anat. Rec.*, 210, 375–383.

Huard, J.M.T. and Schwob, J.E. (1995) Cell cycle of globose basal cells in rat olfactory epithelium. *Dev. Dynam.*, 203, 17–26.

Mackay-Sim, A. and Kittel, P.W. (1991) On the life-span of olfactory receptor neurons. *Eur. J. Neurosci.*, 3, 209–215.

Mattes, R.D., Cowart, B.J., Schiavo, M.A., Arnold, C., Garrison, B., Kare, M.R. and Lowry, L.D. (1990) Dietary evaluation of patients with smell and/or taste disorders. *Am. J. Clin. Nutr.*, 51, 233–240.

Mbiene, J.P. and Farbman, A.I. (1993) Evidence for stimulus access to taste cells and nerves during development: an electron microscope study. *Microsc. Res. Tech.*, 26, 94–105.

Miller, I.L. and Reedy, F.E. (1990) Variation in human taste bud density and taste intensity perception. *Physiol. Behav.*, 47, 1213–1219.

Mombaerts, P., Wang, F., Dulac, C., Chao, S.K., Nemes, A., Mendelsohn, M., Edmondson, J. and Axel, R. (1996) Visualizing an olfactory sensory map. *Cell*, 87, 675–686.

Moran, D.T., Jasef, B.W., Eller, P.M. and Carter Rowley III, J. (1992) Ultrastructural histopathology of human olfactory dysfunction. *Microsc. Res. Tech.*, 23, 103–110.

Moran, D.T., Rowley, J.C., Jafek, B. and Lovell, M.A. (1982) The fine structure of the olfactory mucosa in man. *J. Neurocytol.*, 11, 721–746.

Morrison, E.E. and Costanzo, R.M. (1990) Morphology of the human olfactory epithelium. *J. Comp. Neurol.*, 297, 1–13.

Morrison, E.E. and Costanzo, R.M. (1992) Morphology of olfactory epithelium in humans and other vertebrates. *Microsc. Res. Tech.*, 23, 49–61.

Nakashima, T., Kimmelman, C.P. and Snow, J.P. (1984) Structure of human fetal and adult olfactory neuroepithelium. *Arch. Otolaryngol.*, 110, 641–646.

Rolls, B.J., Rowe, E.A. and Rolls, E.T. (1982) How sensory properties of foods affect human feeding behaviour. *Physiol. Behav.*, 29, 409–417.

Schwob, J.E., Mieleszko Szumowski, K.E. and Stasky, A.A. (1992) Olfactory sensory neurons are trophically dependent on the olfactory bulb for their prolonged survival. *J. Neurosci.*, 12, 3896–3919.

Shaal, B. and Orgeur, P. (1992) Olfaction in utero: can the rodent model be generalized? *Q. J. Exp. Psychol.*, 44B, 245–278.

Silver, W.L., Mason, J.R., Adams, M.A. and Smeraski, C. (1986) Trigeminal chemoreception in the nasal cavity: responses to aliphatic alcohols. *Brain Res.*, 376, 221–229.

Smith, D.V. and Duncan, H.J. (1992) Primary olfactory disorders: anosmia, hyposmia and dysosmia. In Serby, M.J. and Chobor, K.L. (eds), *Science of Olfaction*. Springer-Verlag, NY, pp. 439–466.

Stevens, J. (1989) Food quality reports from non-institutionalized aged. *Ann. NY Acad. Sci.*, 561, 87–93.

Stevenson, R.D. and Allaire, J.H. (1991) The development of normal feeding and swallowing. *Pediatr. Clin. North. Am.*, 38, 1439–1453.

Strachan, D.S. (1994) Histology of the oral mucosa and tonsils. In Avery, J.K. (ed), *Oral Development and Histology*. Thieme Medical Publishers, NY, pp. 298–320.

Trojanowski, J.Q., Newman, P.D., Hill, W.D. and Lee, V.M.-Y. (1991) Human olfactory epithelium in normal aging, Alzheimer's disease, and other neurodegenerative disorders. *J. Comp. Neurol.*, 310, 365–376.

Wysocki, C.J. and Gilbert, A.N. (1989) National geographic smell survey. Effects of age are heterogenous. *Ann. NY Acad. Sci.*, 561, 12–28.

Wysocki, C.J. and Pelchat, M.L. (1993) The effects of aging on the human sense of smell and its relationship to food choice. *Crit. Rev. Food Sci. Nutr.* 33, 63–82.

Yamaguchi, S. (1991) Basic properties of umami and effects on humans. *Physiol. Behav.*, 49, 833–841.

17

FUTURE TECHNOLOGIES ENVISAGED FROM MOLECULAR MECHANISMS OF OLFACTORY PERCEPTION

G.A. BELL

INTRODUCTION

Our search for the mechanisms that explain how the sense of smell works at the molecular level has been greatly advanced in recent years by three areas of discovery. Firstly, we now have a clear, albeit incomplete, understanding of how volatile chemical stimuli (odorants) impact on the olfactory receptor nerve cells in the nose, by way of macromolecules embedded in the nerve membrane. Secondly, we also have new understanding of the mechanisms by which enzymes terminate the effectiveness of the olfactory stimuli. Thirdly, the role of mucosal proteins in transporting odorants through the nasal mucus to the receptor sites on the olfactory cell membranes is being understood in molecular terms.

These new discoveries offer a foundation for a wide-ranging technology based on olfactory macromolecules. In this chapter we will look at what inventions may arise, and which industries they may serve, from understanding these three nasally located sets of mechanisms for the sense of smell.

OLFACTORY TRANSDUCTION

The transduction mechanisms will be the first topic for discussion. It involves the cell membrane-bound receptor proteins, and holds out a promise for future technologies related to selective and highly sensitive sensing, as well as specific fragrance or flavour control.

Transduction is the process by which the events of the physical world become represented as electrical activity in a sensory nerve cell. Early conceptual guidance on the process came from a stereo-chemical model based on the shape of odour molecules and their reception by a matching imprint of their shape on the surface of the receptor nerve cell (Amoore *et al.*, 1964). In these very simple terms, and well in advance of modern receptor biophysics and biochemistry, Amoore *et al.* were remarkable in anticipating, by several decades, the development of answers to the question of how chemical odorants are primarily transduced into nerve signals.

We now know that olfactory transduction mechanisms share many common features with the visual, auditory and vestibular systems of the inner ear (see Torre, *et al.*, 1995).

Like photoreceptors, the olfactory receptor cells work by the action of a stimulus (photon or odorant, respectively) acting on a receptor protein at the receptor cell's surface. Odorants are small molecules, with molecular weights less than 1 kDalton. An odorant is light enough to be breathed into the nose but is sufficiently complex to be recognised as having a structural characteristic unique to that odorant. The recognition is performed by the set of hundreds (possibly thousands) of unique olfactory receptor proteins first described by Buck and Axel (1991) (see review by Sullivan and Dryer, 1995). It is now apparent that a peculiar feature of the odorant (called an epitope) is recognised by a particular genetically unique receptor, while another feature is recognised by a different unique receptor (Ressler *et al.*, 1994). A code, or 'epitope map', analogous to a bank account personal identity number (PIN), forms from the action of the receptor proteins on the cells in which they are embedded. A cascade of biochemical events in each cell follows the binding of the odorant to the receptor (see further information in this volume by Barry and Sullivan). The resulting ionic disturbance on the cell membrane, if sufficiently large, causes a spike discharge (nerve impulse) to travel down the axon of the olfactory cell to its first synapse with the mitral and tufted cells of the olfactory bulb.

In these trains of nerve impulses — travelling from nasal epithelium into the complex neural architecture of olfactory bulb, and beyond to higher destinations in the central nervous system — resides all the basic information the system needs about the concentration and quality of the odorants (Mori and Yoshihara, 1995).

The perceived quality of every odorant begins with this special relationship of molecular recognition between the small airborne odorant and a comparatively large receptor protein (over 40 kDalton), which evolution has provided for its capture. A technology that can harness these receptors can be made to act upon odours of commercial significance — to capture them and so disable them, or to measure their presence and quantity. New molecules could be designed to fit the receptors by mimicking certain odours, but without an odour property.

One such technology would be a sensor made from a particular receptor. It would sensitively and specifically recognise odorants with a specific epitope.

How many different receptors would be needed to encode, fully, a specific odorant molecule? It would depend on the number of epitopes an odorant has. Conceptually, a receptor might have sites for several different epitopes, but this still has to be demonstrated. However, we now have a good indication, based on a combination of evidence, of how we can anticipate the number of receptors involved in encoding the characteristics (or epitopes) of an odorant molecule. The first piece of evidence is that tracing of gene-specific olfactory receptors led to two and only two glomeruli in the olfactory bulb (Mombaerts et al., 1996). (Glomeruli are the ball-shaped entanglements formed by the incoming axons of olfactory nerves from the nose with the dendrites of the mitral cells in the olfactory bulb.)

If a genetically specified receptor macromolecule only ever feeds into two glomeruli (per hemisphere of the brain), then we can get an idea of the number of epitopes an odorant has by looking at the number of glomeruli which the odorant activates. Here we consult a body of evidence already existing in published and unpublished form. This is the second piece of evidence we need: radioactive 2-DG mapping studies of the olfactory glomeruli, which become metabolically active after the nose (usually of the laboratory rat) is stimulated with a single odorant compound (see Stewart et al., 1974; Jourdan et al., 1980; Bell et al., 1987; Bell, 1997; Johnson et al., 1998). By counting the active glomeruli and dividing their number by two, we have an estimate of the number of receptors registering a response to that compound. Each compound has one or more receptors encoding it.

The maps show that odorants vary greatly in their spatial distribution of active glomeruli and in the number activated (see Figure 17.1). Octane, for example, stimulated a small number, two to four, meaning it has only one, perhaps two, receptors. So a single receptor will suffice to capture and identify octane. Making a sensor for octane, using its biological receptor, will therefore be relatively easy. Propionic acid, by comparison, stimulated around 60 glomeruli, so the number of receptors needed to fully encode it is 30, that is 60 divided by two. Making a highly specific sensor for propionic acid from its receptors will not be a simple matter.

Sensors that are truly specific for an odorant will have to possess all the receptors needed to recognise the full complement of epitopes for that odorant. Therefore it seems inevitable that to fulfil the selectivity requirement for a receptor-based sensor system, more than one sensor, that is an array of sensors, will be needed to create the specific code for most compounds.

Odours in everyday life do not usually consist of single compounds. Almost every smell encountered consists of a complex mixture of many, even hundreds, of odorants. At the brain, the olfactory information is transformed into a unified sensory experience, as well as creating and evoking memories of names, places and feelings, etc. We say 'That smells like coffee', or 'That reminds me of grandma's kitchen'. We put a single name or classification onto the complex olfactory data set.

This presents a difficult problem for the sensor array designer, as each 'PIN' must be unambiguously recognisable. Once that is achieved, they can be dealt with by neural nets or other such elegant classification tools (see Levy in this volume). But, before that, the elements of different PINs must not be allowed to become confused.

To arrive at a solution to this problem, the technologist will need to look at how the nose solves the problem: by preconditioning the odorant mixture prior to processing it. In the nose this is done by forcing the odour-laden airstream through a convoluted set of mucus-lined, narrow passages, so that receptors in specific regions of the nose have the best chance of capturing a particular class of compound. It is not yet known on what basis, polarity, lipid solubility, etc., the nose actually sorts the incoming molecules. However, the molecules do appear to arrive at different parts of the receptor cell sheet, and at different times, in order to create a spatiotemporal activation pattern in the nose (see Kauer, 1991; Mackay-Sim and Kesteven, 1994; Youngentob et al., 1995).

The technologist's best strategy will be to harness the receptors for

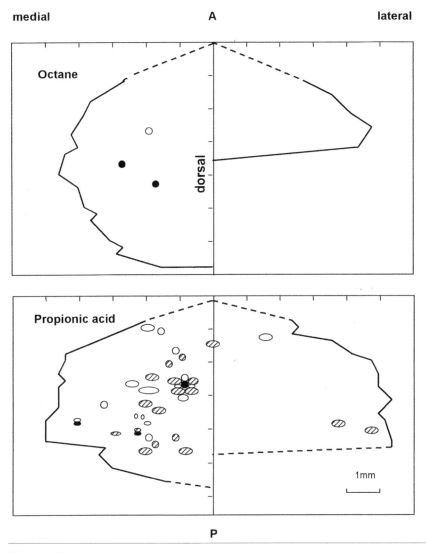

Figure 17.1

Maps of forebrain glomeruli which became metabolically active after stimulation of the rat with single compounds, revealed by the radioactive 2-DG method (Bell, 1997). The glomeruli are not all activated to the same degree. The greatest metabolic activity is shown by black-filled circles, being three times greater than background activity. Hatched areas are two and a half times the background, open circles are twice the background activity level. By examining olfactory bulb sections from which the maps were made and measuring the diameter of the glomeruli, the number of activated glomeruli could be determined. There were 3 glomeruli activated by octane and 60 by propionic acid in the olfactory bulbs shown here. Different levels of metabolic activity in pairs of glomeruli probably indicate variable affinities of receptors for the epitopes on the odorant molecule. A = anterior B = posterior the key odorants in industrially important complex odours, such as nuisance odours from animal farms, taints, leaks in containers and packaging, traces of dangerous substances, or telltale metabolites of bacteria.

Two important demonstrations, from the rat and the nematode, show that odour receptors for specific odorants can ultimately be cloned and farmed for technological purposes (Raming *et al.*, 1993; Sengupta *et al.*, 1996). In addition, studying the interaction between the identified receptor and its ligand (ie the odorant specific to the receptor) will allow the precise nature of the binding and recognition mechanism to be known. From that will follow creation of 'designer' odorants and receptors. For example, a receptor might be structurally manipulated to accept a different, commercially valuable odorant, or a slight modification to the ligand might change it into an odourless obstruction to, or competitor for, the unmodified ligand, and thereby create a potentially valuable deodo-rant. So-called designer odorants might be valuable in turning on, or off, animal behaviour of com-mercial significance, such as appetite for food or sex. Identifying the gene for a specific odorant receptor will also open up possibilities of gene therapy for anosmic people.

The number of genes required to serve a large number of unique receptors is large. The portion of the human genome dedicated to the human olfactory receptors is probably about half a percent of the genome. If these are mapped by the US Human Genome Project, which aims to map the entire human genome, the human olfactory genes will be a rich resource for human olfactory biotechnology. An ensuing gene therapy, for instance, would protect the sense of smell in old age.

Some issues need to be addressed before practical use can be made of a receptor technology. The most pressing issue will probably be their functionality outside the system in which they normally live. Will they survive and perform in isolation from their membranes? Can they be separated from the membrane without damage? Will their functional sites remain accessible to the odorant? Can they be attached and orientated appropriately on an artificial surface? Do they need mucus to survive and function? Will they release the bound odorant within a useful time?

AN ALTERNATIVE TO RECEPTORS: MUCOSAL CARRIER PROTEINS

Olfactory receptor cells have a special environment into which their apical knobs and cilia protrude, namely the olfactory mucus. We now know that the olfactory mucus is a complex substance with a special part to play in delivering the chemical stimuli from the outside world to the cell membranes of the receptor cells (see Pelosi,1996).

Olfactory perception begins with the transit of odorants through the mucus layer covering the olfactory epithelium. During its penetration of the mucus, the odorant takes part in a number of biochemical events before it reaches and binds with the receptor proteins embedded in the membrane of the receptor cell.

All animals seem to require an aqueous interface medium through which the odorant has to move before it reaches the receptor.

The aqueous phase of the mucus is maintained by the binding of water molecules to large glycoproteins, or 'mucins'. Is the sense of smell affected when mucin production changes, or when mucus thickens or dries up, as in sick or aging humans? Understanding the production of mucus might lead to treatment for disease or age-related deficits, and the improvement of quality of life for the aging human population.

There are a number of other important and recent developments in our understanding of the role of olfactory mucus. For example, we know that the structure and elasticity of the mucus is maintained by polymeric peptide chains of 'olfactomedin', a name derived from a role it was thought to have in mediating olfactory transduction (Snyder *et al.*, 1991). These molecules might be usefully cloned as non-specific odorant binding agents for the purpose of deodorising.

Sugars in the mucus might be crucial to embellishing the odorant into a specific complex, which then fits the receptor's active site, or sites. Better understanding of these processes might improve odour sensitivity in future detection devices, and hold the key to how one receptor might accommodate multiple specificities for odorants.

Other proteins in the nasal mucus play a part in odorant recognition, and may also have potential technological value. These will be discussed next: the odorant-binding proteins (OBPs), and the odorant degrading enzymes (ODEs).

ODORANT-BINDING PROTEINS

Olfactory binding proteins (OBPs) resemble carrier proteins found in the liver and excreted in the urine, and belong to the same family of proteins known as lipocalins. They are produced in great quantities at several glandular sites in the nasal cavity. The secretions of these glands are constantly washed away as the mucus moves from the front of the nose to the back, and down the throat. The OBPs are soluble, and of low molecular weight, and each molecule binds several different odorants. They function to carry hydrophobic odorants through the aqueous medium to the receptors. They have been found in high concentration in the nasal mucus of mammals such as cows, pigs, rabbits and mice (Pelosi, 1997).

The presence of OBPs in the vomeronasal organ (VNO) suggests that some forms of OBPs are specific to pheromones. The VNO is a chemosensory organ resembling a small bag with a long throat, which points forward to within the region of the opening of the nasopalatine duct (a tube connecting the nasal cavity and the mouth). It is implicated in reproductive functions in many mammals, and is observable as a vestigial pit in the human nose, but may retain some functionality (Moran *et al.*, 1995). In animals with well-developed VNOs, the OBPs may have the potential for future technology, as will be briefly demonstrated.

The urinary lipocalins, molecules of the OBP family, have been shown to carry pheromones (Bacchini *et al.*, 1992; Robertson *et al.*, 1993), and therefore probably serve to distribute, concentrate and conserve pheromones. This was recently elegantly demonstrated in no less a mammal than the elephant (Rasmussen *et al.*, 1998). The receptive male, investigating a urine source, dribbles nasal secretions down its trunk, which are mixed with the urine before being snorted back up and delivered to the VNO. Thus it takes into its nose the pheromone odorant with its carrier protein conveniently attached, thereby accelerating recognition, or tuning the odorant molecule for a specific recognition function within that animal's species. Technological applications of sexual pheromone-related OBPs, or OBPs involved in other chemical signalling between animals, will follow from better knowledge of these proteins. For example, the appropriate OBP might make a big difference to the efficiency of synthetic boar pheromones now commonly used in the pig industry.

Several types of OBP are heavily concentrated in the lymph of insect olfactory sensilla. They appear to have specificity either for pheromones (PBPs) or for general odorants (GOBPs). A PBP has been successfully expressed in *Escherichia coli* and produced in sufficient quantity to allow X-ray crystallography and NMR studies of the protein (Prestwich,1993), thereby laying the ground-work for future production of OBPs by biotechnology.

The way is therefore opening up to express other OBPs in host cells, and manufacture purified, cloned OBP molecules in quantities suitable for technological applications. As they are more likely to retain functionality outside the molecular environment of the mucus, compared with the membrane-bound receptors, they might have much greater viability in artificial materials and surfaces.

What could one do with them? Facilitating delivery of synthetic pheremones to animals, by carrying the pheromones on their appropriate OBPs, should have immediate commercial importance for agriculture and animal production. They might be used in various forms of insect and animal control, perhaps to dilute or nullify the efficacy of chemical signalling between insects and between other animals. This could lead to eradication of diseases such as sleeping sickness and malaria. They could enhance fragrance or flavour in perfume or food systems. They could be impregnated into fabric and used to deodorise unwanted stench. They might be applied to specifically capture and concentrate certain valuable odours. The technological pay-offs from OBPs lie in their freedom from being membrane-bound, and from their broad-range specificity for classes of odorants.

ODORANT-DEGRADING ENZYMES

Degrading or terminating enzymes (ODEs) are another form of soluble protein found in the olfactory mucosa, whose functions indicate exploitation in future technologies, particularly as catalysts, detoxicants and deodorants.

The olfactory system has an efficient mechanism for clearing away inhaled odorants and pheromones, most of which are toxic to the animal in high quantities. The nose has to continuously detoxify itself, while preparing itself to identify and quantify a fresh whiff of odour. It does so with the aid of several enzymes, of a kind found in similar concentrations in the liver (eg cytochrome P-450). These enzymes degrade the odorants and render them ineffective as chemical stimuli, but play no role in stimulating the receptors. The ODEs may also have the property of amplifying the effective strength of weak odorants, by clearing a path for them, as it were. They may also attenuate strong odorants, by reducing the number of functional odorant molecules, on their path through the mucus to the receptor (Pelosi, 1996).

The degradation enzymes of the vertebrate nose, with the single exception of those found in the pig (Persaud et al., 1988), are not specific to particular odorants. In the insect, however, a number of pheromone-specific degrading enzymes has been found in the olfactory lymph (Vogt and Riddiford,1981; Rybczynski et al.,1989). These specific ODEs may be useful in controlling insect pests.

THE NEXT HORIZON

Understanding the molecular mechanisms of olfaction, from the odorant's entry through the nasal mucus to the termination of their role as chemical stimuli, will provide, in the foreseeable future, new receptor-based devices for chemical analysis, quality control, safety and security. Derived from nature, the receptive molecules will be used in safer and better deodorants. New materials that will embody chemosensory molecules, or imitate their mechanisms, will detect unsafe or unhealthy objects and substances. They will also select and remove toxicants and pollutants from food, water, air and earth. They will be the essential ingredient in electronic control and monitoring devices in manufacture, as well as in making household appliances safer and more efficient. A new generation of diagnostic tools will identify a wide range of illnesses from the whiff of a patient's breath. New kinds of chemosensory pharmaceuticals will find and degrade specific chemical targets for better health and treatment of diseases. Through radically improved ability to control odour, we will create better processes and products, and a safer, healthier, cleaner environment.

New knowledge about brain processing of olfactory information will also lead to useful and valuable technology. Control systems will imitate the olfactory bulb and the integrative areas of the sensory systems, to enhance processing of complex information.

Identification of molecular olfactory mechanisms in recent years has been pivotal in transforming our view of the future of chemosensory technology, and provoking serious discussion and investment in new technologies. The way lies ahead for harnessing the molecular mechanisms of olfactory perception for the public, and commercial, good. In the future, chemosensory technology will be commonplace.

REFERENCES

Amoore, J. E., Johnston, J. W. and Rubin, M. (1964) The stereochemical theory of odor. *Sci. Amer.*, 210, 42–49.

Bacchini, A., Gaetani, E. and Cavaggioni, A. (1992) Pheromone-binding proteins in the mouse *Mus musculus*. *Experientia*, 48, 419–421.

Bell, G. A. (1997) Receptors for odorants of similar and dissimilar qualities and molecular structure visualised by 2–DG maps of olfactory bulb glomeruli. *Chem. Senses*, 22(6), 645–646.

Bell, G. A., Laing, D. and Panhuber, H. (1987) Odour mixture suppression: Evidence for a peripheral mechanism in human and rat. *Brain Res.*, 426, 8–18.

Buck, L. and Axel, R. (1991) A novel multigene family may encode odorant receptors: A molecular basis for odor recognition. *Cell*, 65, 175–187.

Johnson, B.A., Woo, C.C. and Leon, M. (1998) Spatial coding of odorant features in the glomerular layer of the rat olfactory bulb. *J. Comp. Neurol.*, 393, 457–471.

Jourdan, F., Duveau, A., Astic, L. and Holley, A. (1980) Spatial distribution of [^{14}C]2-deoxyglucose uptake in the olfactory bulbs of rats stimulated with two different odours. *Brain Res.*, 188, 139–154.

Kauer, J.S. (1991) Contributions of topography and parallel processing to odor coding in the vertebrate olfactory pathway. *Trends Neurosci.*, 14(2), 79–85.

Mackay–Sim, A. and Kesteven, S. (1994) Topographic patterns of responsiveness to odorants in the rat olfactory epithelium. *J. Neurophysiol.*, 71, 150–160.

Mombaerts, P., Wang, F., Dulac, C., Chao, S.K., Nemes, A., Mendelsohn, M., Edmondson, J. and Axel, R. (1996) Visualizing an olfactory sensory map. *Cell*, 87, 675–86.

Moran, D.T., Monti-Bloch, L., Stensaas, L.J. and Berliner, D.L. (1995) Structure and function of the human vomeronasal organ. In Doty, R.L. (ed), *Handbook of Olfaction and Gustation*. Marcel Dekker, NY, pp. 793–820.

Mori, K. and Yoshihara, Y. (1995) Molecular recognition and olfactory processing in the mammalian olfactory system. *Prog. Neurobiol.*, 45, 585–619.

Pelosi, P. (1996) Perireceptor events in olfaction. *J. Neurobiol.*, 30(1), 3–19.

Persaud, K.C., Pelosi, P. and Dodd, G.H. (1988) Binding and metabolism of the urinous odorant 5á–androsten–3–one in sheep olfactory mucosa. *Chem. Senses.*, 13, 231–245.

Prestwich, G.D. (1993) Bacterial expression and photoaffinity labeling of a pheromone binding protein. *Prot. Sci.*, 2, 420–428.

Raming, K., Krieger, J., Strotman, J., Boekhof, S., Kubick, S., Baumstark, C. and Breer, H. (1993) Cloning and expression of odorant receptors. *Nature*, 361, 353–356.

Rasmussen, L.E.L., Lazar, J., Greenwood, D., Feng, L. and Prestwich, G.D. (1998) Initial characterizations of secreted proteins from Asian elephants that bind the sex pheromone, (Z)-7-dodecenyl acetate. *Chem. Senses*, 23(5), 591.

Ressler, K.J., Sullivan, S.L. and Buck, L.B. (1994) Information coding in the olfactory system: evidence for a stereotyped and highly organized epitope map in the olfactory bulb. *Cell*, 79, 1245–1255.

Robertson, D.H.L., Beynon, R.J. and Evershed, R.P. (1993) Extraction, characterisation and binding analysis of two pheromonally active ligands associated with major urinary protein of the house mouse. (*Mus musculus*). *J. Chem. Ecol.*, 19, 1405–1416.

Rybczynski, R., Regan, J. and Lerner, M.R. (1989) A pheromone-degrading aldehyde oxidase in the antennae of the moth *Manduca sexta*. *J. Neurosci.*, 9, 1341–1353.

Sengupta, P., Chou, J.H. and Bargmann, C.I. (1996) *odr–10* encodes a seven transmembrane receptor required for responses to the odorant diacetyl. *Cell*, 84, 899–909.

Snyder, D.A., Rivers, A.M., Yokoe, H., Menco, B.PH.M. and Anholt, R.R.H. (1991) Olfactomedin: purification, characterization, and localization of a novel olfactory glycoprotein. *Biochem.*, 30, 9143–9153.

Stewart, W.B., Kauer, J.S. and Shepherd, G.M. (1974) Functional organization of rat olfactory bulb analysed by the 2-deoxyglucose method. *J. Comp. Neurol.*, 185, 715–734.

Sullivan, S.L. and Dryer, L. (1995) Information processing in mammalian olfactory system. *J. Neurobiol.*, 30(1), 20–36.

Torre, V., Ashmore, J.F., Lamb, T.D. and Menini, A. (1995) Transduction and adaptation in sensory receptor cells. *J. Neurosci.*, 15(12), 7757–7768.

Vogt, R. G. and Riddiford, L.M. (1981) Pheromone binding and inactivation by moth antennae. *Nature*, 293, 161–163.

Youngentob, S.L., Kent, P. F., Sheehe, P.R., Schwob, J. E. and Tzoumaka, E. (1995) Mucosal inherent activity patterns in the rat: evidence from voltage sensitive dyes. *J. Neurophysiol.*, 73(1), 387–398.

18

ELECTRONIC SENSOR TECHNOLOGIES FOR THE FOOD AND ALLIED INDUSTRIES

A. MACKAY-SIM

THE MAMMALIAN NOSE

There can be no doubt that a fuller understanding of the olfactory system will provide new ideas and new materials for the development of a 'bionic nose' to rival our own remarkable sense organ. The human sense of smell can detect and discriminate thousands of different odours (at concentrations as low as parts per trillion for some), from the fragrances of food and flowers to the smells of sweat and the seaside. Despite the importance of smells in our lives it is only recently that there has been any electronic means to identify and quantify smells in the air. Even now the most reliable tool is still the human nose.

Although the human nose is highly sensitive to individual odours, it is not very good at the analysis of mixtures. Untrained observers have difficulty identifying correctly the components of mixtures of just two or three compounds. Even perfumers and flavourists have difficulty in identifying more than six components of a complex, nineteen component mixture (Laing and Francis, 1989). Yet odours of food and beverages, environmental odours, pleasant or offensive odours — virtually all odours — are complex mixtures of many components. Rarely is an odour molecule encountered alone.

Development of electronic odour sensors with superior discriminative power would therefore have extensive application in industrial processing and in environmental monitoring. The most useful sensors would be of two types: 'single odour' sensors, which would be selectively sensitive to one target odour in a mixture (for the detection of contaminants, for example) and 'mixture analysers' which would be able to quantify numerous components in complex mixtures of odours (for example, analysis of foods and beverages during processing).

The design of any odour sensor, electronic or biological, depends on the interaction between the odour molecule and a receptor molecule on the surface of the sensor element or sensory nerve cell. The response of the sensing system is a compromise between sensitivity, selectivity and reusability. Very high affinity binding of an odour to a sensor may give high sensitivity and selectivity, but the reaction may not be reversible, so that the sensor could only be used once. In the nose, the sensory nerve cells are not very selective for individual molecules, but are responsive to a wide range of related molecules. The sensory cell response is reversible and the high sensitivity of the system arises from the very large number of sensory cells. There are thousands of sensory cells with similar selectivity whose output converges onto single nerve cells in the brain. This convergence greatly amplifies the signal and helps to compensate for lower sensitivity of the sensory cells in the nose.

The sense of smell is thus based on millions of sensory cells, each of which responds to many odour molecules. Within the nose these cells are arranged so that cells of similar responsiveness lie close to each other in the olfactory epithelium (the sense organ of smell). Distant regions of the epithelium have dissimilar properties (Mackay-Sim and Kesteven, 1992). Different odours therefore produce different 'maps' of responses from across the epithelial surface, each odour-evoked map being unique (Mackay-Sim and Kesteven, 1992). These response maps are conveyed to the brain where they are interpreted as different odours.

At the surface of the sensory cells, odorous molecules bind with 'odour-sensing' receptor proteins on the surface of the sensory cell. There are many hundreds of these receptor proteins and each sensory cell probably has only one or two types of receptor molecules (Ressler *et al.*, 1993). Initial data on the odour-binding properties of one receptor protein suggested that it was broadly selective (Raming *et al.*, 1993), but a recent study of another showed high selectivity to a small number of aliphatic alcohols, with maximal sensitivity to octanol (Zhao *et al.*, 1998). Receptor proteins may thus vary in their selectivity and sensitivity. The electrical signals of thousands of cells with the same receptor molecules converge on a few cells in the brain, such that each of these brain cells is highly sensitive to discrete aspects of the chemistry of each odour, such as its carbon-chain length, or the presence of a particular reactive group such as alcohol or ketone (Mori and Yoshihara, 1995).

The discriminative power of the olfactory system probably depends on the time-dependent variability of the electrical signals from the sensory cells (Laurent, 1997). Different odours may thus interact with the same receptor proteins on the same sensory cells but with different time-dependent kinetics. This can result in different time-dependencies in the signals from the sensory cells to the brain. There is another inherent time-dependency in the olfactory signals to the brain, which is imposed by breathing and sniffing and complicated by the physico-chemical interactions of odours with the mucus in the nasal cavity (Kent *et al.*, 1996).

From the accumulated physiological and anatomical data, it seems that the olfactory system uses a combination of both spatial and temporal information to discriminate odours. When similar principles of spatiotemporal patterning are designed into an electronic odour-sensing sensor system, the discriminative power of the system increases significantly (Mackay-Sim *et al.*, 1993; Saunders *et al.*, 1995a; Dickinson *et al.*, 1996).

ELECTRONIC NOSES

An electronic odour sensor is essentially comprised of two parts: reactive surfaces with which odour molecules interact, and some electronic means to detect and identify that interaction. Some common sensor systems are based on conducting polymers, tin oxide semiconductors, optical fluorescence and piezoelectric devices.

Conducting polymers are a new range of plastics whose conductance changes, depending on the nature of the side chains along the polymer backbone. In addition, there can be further conductance changes when the side chains interact with odours. Typical odour detection systems use arrays of sensors, each with a different polymer, giving different conductance changes when stimulated with odours (Persaud *et al.*, 1996). One of the commercially available odour sensor systems is based on conducting polymers.

Tin oxide semiconductors rely on the ionisation of odour molecules when they are raised to high temperatures. The sensitivity of the sensors depends on the oxidation state of the tin oxides used, and their electrochemical interaction with the ionised odour molecules. There are at least two commercially available odour-sensing systems based on tin oxide semiconductors.

Another interesting method for odour-sensing arrays uses the optical detection of fluorescence changes when odours interact with the coated surface of an optical fibre (Ingersoll and Bright, 1997). The coated end of the optical fibre is illuminated via the fibre, and changes its fluorescence in the presence of odours. Different coatings provide the selectivity, and the emitted fluorescence is detected at the other end of the optical fibre.

Historically, the most popular electronic detection system for odour sensing has been the piezoelectric crystal oscillator, whose frequency of oscillation varies with the mass of substances on its surface. There have been several commercial developments using the piezoelectric sensor. See Barnett (this volume) for more details and illustrations of these sensing devices.

ELECTRODE DESIGN

The chemistry of the sensor surface is critical for the selectivity and sensitivity of the sensor, especially when trying to design 'single chemical' sensors for use in contaminant detection or clinical analyses. This high specificity has been obtained with the use of biological recognition molecules, such as enzymes and antibodies. Enzymes that produce reactive oxygen, which is then detected amperometrically, can be used. Antibodies have been used on piezoelectric devices and optical fluorescence devices.

Unfortunately, the use of enzymes and antibodies poses problems of stability and longevity, and their potential in commercial devices has not been realised. An alternative to these biological recognition molecules is emerging in a new field of synthetic chemistry, 'molecular imprinting'. Molecular imprinting is a method of producing highly stable plastic polymers which contain within their chemical matrix highly selective sites which bind with high affinity to molecules of interest (Kriz *et al.*, 1997). When the polymer is made, it is assembled around a target molecule. Once the polymer is formed, the target molecule is extracted from the polymer, leaving sites within it, which will bind selectively with the target molecule. This technology has the potential to rival biological recognition systems for selectivity, but with much greater stability and longevity, and with the ability for mass manufacture. The main problems to be solved for this technology are the long response times (15–60 min), and the fact that polymers with faster responses may compromise selectivity (Kriz *et al.*, 1997).

In most fields of sensor design there has been an acknowledgment that absolute specificity is hard to obtain for a single sensor. The problem is that odour sensors must be reuseable, and so the odour molecule must bind specifically but reversibly. The chemistry of odour-surface interaction is such that reversibility is sacrificed for greater specificity. As a consequence of the limits imposed by surface chemistry, most modern sensor designs capitalise on the lack of specificity by using an array of sensors with differential and partially overlapping sensitivity. In this way, identification of individual odorants relies on the fact that each provides a unique pattern of responses across the array, which can be identified using neural network computing.

Neural network computing is a method of computation based loosely on knowledge of the 'wiring' and operation of nerve cell networks in the brain. Part of the interest in these artificial neural networks lies in their ability to 'learn' or improve their performance of a given task (eg, odour identification) over subsequent 'training' trials. Once a task is learned, the main attribute of these networks emerges, namely, resistance to 'background interference' or incomplete data (generated, for example, by the presence of other components in a mixture). Most of the advanced odour sensor instruments currently on the market employ artificial neural networks. Operationally, the main drawback of these sensors is that their performance can be limited by the time it takes to train the systems to discriminate the odours of interest. Once learned, though, all subsequent discriminations can be achieved very quickly.

The goal of our research is to develop electronic odour detection systems modelled on the mammalian olfactory system, using piezoelectric crystal sensors and neural network computing. A future goal is to use olfactory receptor proteins as the sensor elements.

TIME-DEPENDENT RESPONSES OF PIEZOELECTRIC SENSORS

Chemically modified quartz crystals have been used for many years for the detection of gases. Chemists have used chromatography chemistry to customise piezoelectric oscillator surface waves, and thereby to devise a variety of oscillators sensitive to particular gases.

Despite the potential applications of such devices for the detection of odours, there are few available commercially. One reason for this is the variability of the response of these devices, which is affected by temperature, humidity, pressure and flow rate. We developed a testing system with which all these variables can be controlled and varied independently, using design principles developed to study the sense of smell (Mackay-Sim *et al.*, 1993).

With this system we demonstrated that the responses of individual sensors were actually very repeatable, but that different sensors varied enormously when tested under identical conditions (Mackay-Sim *et al.*, 1993). Another important variable in the responses of these sensors is the relative viscosity of the chemicals used to treat the sensors to make them sensitive to different odours (James *et al.*, 1994). The viscosity of the coating changes with temperature, and probably even changes in the presence of some odours. These observations demonstrated that it may be very difficult to manufacture sensors with repeatable odour responses, making it necessary to calibrate each sensor individually.

The usual indicator of sensor response is the maximum change in oscillation frequency when it is exposed to an odour. It is this response which is so variable between odours. Fortunately, our controlled testing system revealed an important new measurement of sensor response which identifies each odour molecule almost uniquely: the rate of change of the sensor output has a characteristic shape, indicating, in many cases, a complex bonding process between the sensor and the odour. We have called this response the 'kinetic signature'. The major advantage of this measurement is that is repeatable. All sensors with the same surface coating give exactly the same response to the same odour (Saunders *et al.*, 1995a).

In order to take advantage of the kinetic signature, the electronic odour sensor system must present the odours to the sensor in a timed pulse of air, similar to a natural sniff. Thus, an electronic nose must include not only the detection system, but also a method of delivering a controlled burst of air to the detection system. In addition to temporal control of the stimulus, and measurement of the time dependent response, an electronic nose which mimics the natural olfactory system must also include spatial information about the location of the sensing elements. This can be done using a new modification of neural network computing.

SPATIAL CODING IN SENSOR ARRAYS USING TIERED ARTIFICIAL NEURAL NETWORKS

Using the biology of the nose as a guide, we have developed a new type of neural network computing which generates a three dimensional odour response map for the recognition of odours (Saunders et al., 1995b). This system uses an array of piezoelectric sensors whose responses are fed into an artificial neural network, whose output is a two dimensional array of identified odours on one axis, and sensors on the other axis. The magnitude of the network response provides the third axis. When kinetic signatures are used as the input information into this artificial neural network, the system becomes very accurate in identifying odours. Using a six sensor array, we trained the system to discriminate among 18 odours. After training, it reliably discriminated the odours from each other, including adjacent members of a series of similar acetates and amines.

Our experiments have shown that an electronic nose, designed to incorporate the principal features of the natural olfactory system, is able to discriminate odours very reliably, and better than systems which do not include spatiotemporal features of the sensor output. A similar conclusion was subsequently drawn from experiments using arrays of optical fluorescence sensors, in a sensor system designed along similar principles (Dickinson et al., 1996). The next step towards improving the performance of the electronic nose might be to use the natural biological sensing element, the olfactory receptor proteins, as the surface coating of the electronic nose.

RECEPTOR PROTEINS FOR SENSOR COATINGS?

Piezoelectric sensors are selectively sensitive to different odours, or groups of odours, because they are coated with different substrates which react reversibly and selectively with those odours. The substrates used have usually been organic polymers of various types, but in several cases protein antibodies have been used in order to target sensitivity to molecules of interest. Another possibility is to provide biological specificity for odour sensing by using olfactory receptor proteins as surface coatings. It is estimated that there may be as many as one thousand of these proteins, each of which is presumed to have slightly differing sensitivity to odours (Raming et al., 1993; Ressler et al., 1993; Zhao et al., 1998). It is thought that the olfactory receptor genes may each be responsive to some small physico-chemical aspect of each odour molecule. In this way the sensory cells in the nose probably act as detectors of 'features' of odour molecules, rather than detectors of whole molecules (Turin, 1996). The odour receptor proteins thus represent a valuable potential resource of recognition molecules, which have been shaped by millions of years of evolution to detect physico-chemical aspects of odour molecules. If they could be isolated they would have great potential as the odour-sensing elements in electronic noses.

Over the last few years we have been exploring ways in which large quantities of olfactory receptor proteins could be manufactured. One way to do this would be to manufacture them in tissue culture. This has been tried once, with limited success, by genetically engineering a commonly used, but non-olfactory cell type (Raming *et al.*, 1993). One of the difficulties with that approach is that even though a gene may be present, there are many steps to the production of a functional protein: the gene must be correctly transcribed into mRNA; the mRNA must be translated into the correct sequence of amino acids, which must be folded into the correct shape to make a protein; finally the protein must undergo biochemical modification to make it functional, and locate it in the outer membrane of the cell.

Some of these steps are dependent on the type of cell being used to manufacture the protein, that is, the full sequence of events may not occur in all cell types. For this reason we are attempting to express receptor proteins in more appropriate cells, namely, genetically engineered cells from the olfactory epithelium. These cells were genetically engineered so that they are 'immortalised', genetically homogeneous, and can grow continuously in culture (MacDonald *et al.*, 1996). We are now exploring ways in which to engineer these cells to produce receptor proteins. In essence we are inserting DNA-encoding receptor proteins into these cells, and testing whether they produce an active protein by stimulating them with odours, and recording appropriate physiological responses from them. If this route is successful there is great potential for isolating the active receptor proteins, and incorporating them as the sensing elements in arrays of biosensors.

ODOUR-SENSING ROBOTS

Although many applications of odour sensing might require arrays of sensors which can discriminate among many different odours, or assess the complex aroma of a mixture of odours, some odour sensors might need to be targeted to a single odour only. Such applications could be the identification of the presence of a contaminating odour in food processing, or the location of a gas leak along a

Plate 4

A robot which follows an odour trail. Odours are detected by the sensor (at the 'front' end of the robot) and the sensor electronics. The mini-robot is powered by the onboard battery, which drives the electronics and the motors on the wheels. In the foreground is shown a $2 coin for scale. This robot was made by Dr Andrew Russell, Monash University.

production line (see Plate 4). Similarly, environmental monitoring of odorous pollution from a manufacturing site may require sensors with only a limited range of sensitivity.

One application for single odour sensor technology is robot guidance. We are developing odour guidance systems so that robots can follow odour trails (Deveza *et al.*, 1994). One application for this technology is to guide robots through a space without the need for sophisticated machine vision. The trail can simply be laid so as to avoid whatever objects are present at the time. The trail is volatile, so it has a limited life and can be changed when conditions dictate. Another obvious application for this technology is for finding the location of gaseous leaks (Russell *et al.*, 1994). In this case an odour-guided robot follows an odour plume upwind using a zigzag strategy based on animal strategies (Russell *et al.*, 1994; Consi *et al.*, 1995).

CONCLUSION

The goal of our research is to develop electronic odour sensors for a number of different applications. We have already developed systems for applications such as environmental monitoring and robot guidance, which require only simple sensors that can monitor single odours or single categories of odours. We have simple sensors which can be used as input to data loggers in 'odour surveillance' applications. We have multi-sensor arrays using 'kinetic signatures', and a newly developed artificial neural network which has the ability to learn 'on the job', and to signal when one of its sensor elements is no longer functional. These developments in electronics, computing and chemistry have continued in parallel with the construction and characterisation of the olfactory sensory cell lines. Our biggest challenge is now before us: to combine biology and engineering in a truly bionic nose. We imagine a multi-sensor array which incorporates olfactory receptor proteins, and an artificial neural network, which will be able to make the subtle assessments of the human nose, but be better at identifying the components in a complex mixture of odours.

ACKNOWLEDGMENTS

This work was supported by grants from the Australian Research Council.

REFERENCES

Consi, T.R., Grasso, F., Mountain, D., Atema, J. (1995) Exploration of turbulent odor plumes with an autonomous underwater robot. *Biol. Bull.*, 189, 231–232.

Deveza, R., Russell, A., Thiel, D.V., Mackay-Sim, A. (1994) Odour sensing for robot guidance. *Int. J. Robotics Res.*, 13, 232–239.

Dickinson, T.A., White, J., Kauer, J.S., Walt, D.R. (1996) A chemical-detecting system based on a cross-reactive optical sensor array. *Nature*, 382, 697–700.

Ingersoll, C.M., Bright, F.V., (1997) Using fluorescence to probe biosensor interfacial dynamics. *Anal. Chem.*, 69(13), A403–A408.

James, D., Thiel, D.V., Bushell, G.R., Busfield, K.W., Mackay-Sim, A., (1994) Phase change and viscosity effects on a quartz crystal microbalance. *Analyst*, 119, 2005–2007.

Kent, P.F., Mozell, M.M., Murphy, S.J., Hornung, D.E. (1996) The interaction of imposed and inherent olfactory mucosal activity patterns and their composite representation in a mammalian species using voltage-sensitive dyes. *J. Neurosci.*, 16, 345–353.

Kriz, D., Ramström, O., Mosbach, K. (1997) Molecular imprinting: new possibilities for sensor technology, *Anal. Chem.*, 69(11), A345–A349.

Laing, D.G., Francis, G.W. (1989) The capacity of humans to identify odours in mixtures. *Physiol. Behav.*, 46, 809–814.

Laurent, G. (1997) Olfactory processing: maps, time and codes. *Curr. Opin. Neurobiol.*, 7, 547–553.

MacDonald, K.P.A., Mackay-Sim, A., Bushell, G.R., Bartlett, P. (1996) Olfactory neuronal cell lines generated by retroviral insertion of the *n-myc* oncogene display different developmental phenotypes. *J. Neurosci. Res.*, 45, 237–247.

Mackay-Sim, A., Kennedy, T.R., Bushell, G.R., Thiel, D.V. (1993) Sources of variability arising in piezoelectric odorant sensors. *Analyst*, 118, 1393–1398.

Mackay-Sim, A., Kesteven, S. (1994) Topographic patterns of responsiveness to odorants in the rat olfactory epithelium. *J. Neurophysiol.*, 71, 150–160.

Mori, K., Yoshihara, Y. (1995) Molecular recognition and olfactory processing in the mammalian olfactory system. *Prog. Neurobiol.*, 45, 585–619.

Persaud, K.C., Khaffaf, S.M., Hobbs, P.J., Sneath, R.W. (1996) Assessment of conducting polymer odour sensors for agricultural malodour measurements. *Chem. Senses*, 21, 495–505.

Raming, K., Krieger, J., Strotmann, J., Boekhoff, I., Kubick, S., Baumstark, C., Breer, H. (1993) Cloning and expression of odorant receptors. *Nature* 361, 353–356.

Ressler, K.J., Sullivan, S.L., Buck, L.B. (1993) A zonal organization of odorant receptor gene expression in the olfactory epithelium. *Cell*, 73, 597–609.

Russell, A., Thiel, D., Deveza, R., Mackay-Sim, A. (1994) A robotic system to locate hazardous leaks. In *IEEE International Conference on Robotics and Automation*, IEEE Computer Society Press, Los Alamitos, CA, pp. 2672–2677.

Saunders, B.W., Thiel, D.V., Mackay-Sim, A. (1995a) Response kinetics of chemically-modified quartz piezoelectric crystals during odorant stimulation. *Analyst*, 120, 1013–1018.

Saunders, B.W., Thiel, D.V., Mackay-Sim, A. (1995b) Olfactory imaging: An electronic nose using tiered artificial neural networks and quartz piezoelectric gas sensors. *Australian J. Intelligent Information Processing Systems*, 2, 1–8.

Turin, L. (1996) A spectroscopic mechanism for primary olfactory reception. *Chem. Senses*, 21, 773–791.

Zhao, H., Ivic, L., Otaki, J.M., Hashimoto, M., Mikoshiba, K., Firestein, S. (1998) Functional expression of a mammalian odorant receptor. *Science*, 279, 237–242.

MARINE CHEMICAL SIGNALS: DISPERSAL, EDDY CHEMOTAXIS, URINE PHEROMONES, AND THE DEVELOPMENT OF A CHEMOTACTIC ROBOT

J. ATEMA

INTRODUCTION

Among the tasks of the sensory systems are signal identification and movement guidance. Many organisms depend on chemical signals for survival. In particular, aquatic animals compete for food and mates in a sea of chemical information that has been relatively unexplored by us. In the aquatic environment, as in others, signal identification relies primarily on recognition of odour composition and guidance is based on the dynamics of odour dispersal (Atema, 1995). Lobsters, use urine pheromones for chemical communication and social organisation, and mixtures of amino acids and other compounds for food recognition. They have provided the inspiration for our engineering approaches to developing an autonomous underwater robot with high-resolution sensors to test chemosensory navigation and chemical target localisation.

EDDY CHEMOTAXIS

Many natural odour dispersal patterns can be classified as jets or wakes. We chose the common jet model for detailed description as it has both biological relevance and a long history of analysis of fluid dynamics. Typically, the energy of continuous jet injection into a coaxial laminar background flow, such as sea water, creates a turbulent plume of concentration patches that continue to break up into ever smaller patches until viscosity absorbs the last remaining energy. A very small sensor passing through such a plume will record the variously sized patches as a temporal pattern of concentration peaks (see Figure 19.1). Various characteristics of these peaks show predictable spatial gradients and may hold the key to efficient movement based on chemical gradients (chemotaxis). This particular form of chemotaxis may be called eddy chemotaxis to differentiate it from classical chemotaxis based on mean concentration gradients (Atema, 1996).

To measure the dynamics of underwater odour dispersal, we developed an artificial sensor that can resolve odour concentration with the same spatio-temporal bandwidth as the lobster, *Homarus americanus*, an animal that has guided most of our work on marine chemical signals (Atema and Voigt, 1995). The essential anatomical unit for lobster chemotaxis is the aesthetasc sensillum of its lateral

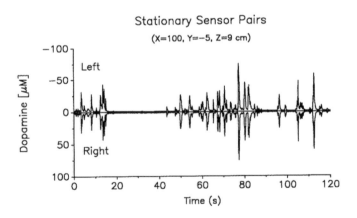

Figure 19.1

Typical peak structure of continuous odour (dopamine) concentration patterns measured with two stationary electrochemical sensors (25 μm tip diameter, 10 Hz sampling rate) spaced 3 cm apart at 9 cm above the ground, similar to bilateral lobster antennules. The sensors are located 100 cm down-current, 5 cm to the right of the centre (X-) axis of the flume. Right side inverted for clarity. (Atema, 1996)

antennular flagellum. Our carbon-filled glass microelectrode matches the aesthetasc diameter of 25 μm (see Plate 5). Temporal resolution of the electrochemical detection process is 10–200 Hz, exceeding the lobster chemosensory flicker-fusion frequency of 5 Hz. To avoid interference by common amines and amino acids in sea water we use the neurotransmitter dopamine as a tracer for odour dispersal. This allows us to describe the fine structure of odour plumes generated under known dispersal conditions.

Plate 5
Glass electrode odour sensor, here placed horizontally near the tips of olfactory (aesthetasc) sensilla (rows of slanted, transparent hairs) of the lobster's lateral antennular flagellum. The electrochemical sensor, size-scaled to one lobster aesthetasc sensillum for comparable spatial resolution , detects dopamine used as a tracer for odour. Temporal resolution of sensor up to 200 Hz; of lobster olfactory receptor cells 1–5 Hz. (Atema)

Lobsters use input from bilateral antennule for efficient direction choice and orientation. With only one antennule they can still locate an odour source, but the process takes much more time and the path is convoluted (Devine and Atema, 1982). To explore the physical basis for this behaviour we measured odour dispersal patterns in a standard jet plume with a pair of aesthetasc-sized electrodes spaced 3 cm apart, matching lobster antennule spacing (Moore *et al.*, 1991;

Moore and Atema, 1991; Grasso *et al.*, in preparation). Three important results emerged (see Atema, 1996, Figure 4):

1 Bilateral comparison of individual peak arrivals can show quickly (~ 1 s) which electrode is located closer to the plume's centre axis, generally the one first hit by the odour patch.

2 The rising concentration (or onset) slopes of individual peaks show a spatial gradient increasing toward the source of odour release.

3 Different plume areas are characterised by typical distributions of peak shapes, creating an odour landscape that, theoretically, could be used for navigation.

Thus, we can extract spatial information from odour-patch distributions in a turbulent jet and perform eddy chemotaxis. This is useful for the design of algorithms to steer underwater robots to sources of chemical signals. However, it does not demonstrate that animals use this information, even though we know that lobsters locating odour sources with bilateral sensors mounted in their olfactory sampling area encounter peak patterns similar to the ones we measure with stationary electrode pairs (Basil and Atema, 1994).

Neurophysiological analysis of responses of lobster olfactory receptor cells to carefully controlled odour pulse shapes and repetition rates has shown that these cells integrate over 0.2 s, fully adapt in ~ 1–3 s, recover partially in ~ 1–5 s and fully in ~ 15–25 s; their flicker fusion frequency varies from 1 Hz to more than 5 Hz (Gomez and Atema, 1996). In addition, background stimuli shift their response functions to higher concentrations (Borroni and Atema, 1988) and repeated stimuli result in cumulative adaptation (Voigt and Atema, 1990). These dynamic characteristics vary from cell to cell, thus laying the foundation for dynamic pattern analysis by the brain. Moreover, the bandwidth properties of the cells favour information processing of fast events matching onset times of common odour pulses and rejecting slower concentration changes. It is tempting to infer that these physiological filter properties evolved to exploit the most useful information contained in turbulent odour plumes.

Such inferences are now being tested directly with both real and robotic lobsters. Real lobsters have been equipped with headgear delivering odour pulses to bilateral antennules which has resulted in turning behaviour (Guenther *et al.*, 1996). Eventually we may be able to create a 'virtual reality' for the lobster and steer it to an imaginary odour source. We are testing the limits of eddy chemotaxis with a robot model, 'RoboLobster' (see Plate 6), built on the known principles of lobster chemical sensing and locomotion (Consi *et al.*, 1994; Grasso *et al.*, 1996). Because we can choose to provide the robot with either chemical or hydrodynamic information or both, we can test the potential for pure chemotaxis and pure rheotaxis (movement determined by water velocity gradient) under different plume conditions such as those found in nature. Whereas it now seems plausible that various animals use eddy chemotaxis/rheotaxis, it seems unlikely that humans have this capability.

Plate 6
RoboLobster is an Autonomous Underwater Vehicle (AUV) designed to mimic a lobster in forward speed and manoeuvrability and steered by bilateral sensors with the purpose of understanding a lobster's capability of following an odour plume to its source. The plume tracking algorithms that result will be used for real world problems of locating underwater sources of chemical dispersal including pollution and explosives. Depicted is the roughly 25 cm long machine facing with its two sensors in a small dye plume (for illustration purposes). Sensor input is evaluated by an on-board computer to make steering decisions that regulate motor speed of two independently controlled wheels. (Tom Consi)

URINE PHEROMONES AND CHEMICAL COMMUNICATION

Lobsters generate three different 'information' currents; each is implicated in chemical communication. A powerful gill current jets forward from bilateral nozzles and extends up to seven body lengths (see Plate 7). This current carries gill metabolites. Urine can be released into this current from bilateral bladders through openings called nephrophores. Histological evidence suggests that the (glyco-proteinaceous) product of a specialised gland in the nephrophore area can be released into the urine, making it a pheromone candidate (Bushmann and Atema, 1996). The lobster controls both the current and urine release in different behavioural situations.

Male lobsters establish a dominance based on a series of initial fights. Losers remember the urine signal of the individual which defeated them (Karavanich and Atema, 1998a, b). We have used chronic catheters to collect the urine and to prevent its release into the environment (Breithaupt *et al.*, 1999; see Plate 8). Individual recognition is lost when urine release is so blocked, and also when olfaction is blocked. Memory of the urine odour is retained for at least one week. Female lobsters prefer dominant males and make their choice from a distance based on male urine cues (Bushmann and Atema, in press).

In nature, social groups appear dominated by a single male who cohabits sequentially with the females, each pairing lasting about ten days. During their nocturnal activities lobsters appear to patrol a small residence area in which all residents know each other and the various lobster shelters. This social system is most likely based on chemical recognition and space memory which is reinforced daily by visiting several shelters each night.

Plate 7
Normal gill current of lobster (Homarus americanus) visualised with dye shows typical, turbulent jet plume. (Linda Golder, Jelle Atema)

Plate 8
Chronic catheter device to collect urine from unrestrained lobsters. Tubes glued around the nephropores lead to a collection device floating on the water surface. (Dan Lindstrom)

CONCLUSION

Practical applications of this work are evident in sensor development, underwater robotics, and identification of chemical compounds. Electrochemical sensing can be developed in areas of product testing and pollution detection. A chemotactic robot can be developed to locate underwater sources of odour, including polluting sources, lost items, and military targets. Pheromones, such as the urine compounds used by lobsters, can be useful tools in stock management and aquaculture; in addition, novel chemical compounds can provide a blueprint for pharmaceutical exploration.

ACKNOWLEDGMENT

This research is financially supported by several grants from the US National Science Foundation, the US National Institute of Health and the Office of Naval Research.

Collaborating investigators include: Drs Jennifer Basil, Paul Bushmann, Thomas Consi, George Gomez, Frank Grasso, Christa Karavanich, Paul Moore, David Mountain and Rainer Voigt.

Photos by Jelle Atema and Linda Golder (Woods Hole, USA); Tom Consi (MIT); Dan Lindstrom, (Gordon College, USA)

REFERENCES

Atema, J. (1995) Chemical signals in the marine environment: dispersal, detection, and temporal signal analysis. *Proc. Natl. Acad. Sci. USA*, 92, 62–66.

Atema, J. (1996) Eddy chemotaxis and odour landscapes: Exploration of nature with animal sensors. *Biol. Bull.*, 191, 129–138.

Atema, J. and Voigt, R. (1995) Sensory biology and behaviour. In Factor, J. (ed) *The Biology of the Lobster*, Homarus americanus. Academic Press, NY, pp. 313–348.

Basil, J. and Atema, J. (1994) Lobster orientation in turbulent odour plumes: Simultaneous measurement of tracking behaviour and temporal odour patterns. *Biol. Bull. (Woods Hole)*, 187, 272–273.

Borroni, P. and Atema, J. (1988) Adaptation in chemoreceptor cells: I. Self-adapting backgrounds determine threshold and cause parallel shift of response function. *J. Comp. Physiol. A*, 164, 67–74.

Breithaupt, T., Lindstrom, D.P. and Atema, J. (1999) Urine release in freely moving catherized lobsters (*Homarus americanus*) with references to feeding and social activities. *J. Exp. Biol.*, 202, 837–844.

Bushmann, P.J. and Atema, J. (1996) Nephrophore rosette glands of the lobster, *Homarus americanus*: Possible sources of urine pheromones. *J. Crust. Biol.*, 16, 221–231.

Bushmann, P. and Atema, J. (1998) Chemically-mediated mate location and evaluation in the lobster, *Homarus americanus*. *J. Chem. Ecol.* (in press).

Consi, T.R., Atema, J., Goudey, C.A., Cho, J. and Chryssostomidis, C. (1994) AUV guidance with chemical signals. *Proc. IEEE Symposium on Autonomous Underwater Vehicle Technology*, Cambridge, MA, 19–20 July, 1994, 450–455.

Devine, D.V. and Atema, J. (1982) Function of chemoreceptor organs in spatial orientation of the lobster, *Homarus americanus*: Differences and overlap. *Biol. Bull.*, 163, 144–153.

Gomez, G. and Atema, J. (1996) Temporal resolution in olfaction: Stimulus integration time of lobster chemoreceptor cells. *J. Exp. Biol.*, 199, 1771–1779.

Grasso, F., Consi, T., Mountain, D. and Atema, J. (1996) Locating odour sources in turbulence with a lobster inspired robot. In Maes, P., Mataric, M.J., Meyer, J.-A., Pollack, J. and Wilson, S.W. (eds), *From Animals to Animats 4: Proceeding of the 4th International Conference on Simulation of Adaptive Behaviour.* MIT Press, Cambridge, MA, pp. 104–112.

Guenther, C., Miller, H., Basil, J. and Atema, J. (1996) Orientation behaviour of the lobster: Responses to directional chemical and hydrodynamic stimulation of the antennules. *Biol. Bull.*, 191, 310–311.

Karavanich, C. and Atema, J. (1998) Individual recognition and memory in lobster (*Homarus americanus*) dominance. *Anim. Behav.*, 56, 1553–1560.

Karavanich, C. and Atema, J. (1998) Olfactory recognition of urine signals in dominance fights between male lobster, *Homarus americanus*. *Behaviour*, 135, 719–730.

Moore, P.A. and Atema, J. (1991) Spatial information contained in three-dimensional fine structure of an aquatic odour plume. *Biol. Bull.*, 181, 408–418.

Moore, P.A., Scholz, N. and Atema, J. (1991) Chemical orientation of lobsters, *Homarus americanus*, in turbulent odour plumes. *J. Chem. Ecol.*, 17, 1293–1307.

Voigt, R. and Atema, J. (1990) Adaption in chemoreceptor cells: III. Effects of cumulative adaptation. *J. Comp. Physiol.*, A166, 865–874.

ELECTRONIC NOSES FOR SENSING AND ANALYSING INDUSTRIAL CHEMICALS

D.B. HIBBERT

INTRODUCTION

Research into the 'bionic nose', or 'electronic nose', is now a maturing area of endeavour — with advances in sensor design and chemistry, the range of applications and the number of available products on the market (Gardener and Bartlett, 1994; Hodgins and Simmonds, 1995). Sensors that rely on the chemical properties of the target molecule, whether it can adsorb at a particular surface, or be oxidised or reduced, have been developed for a variety of analytes. Popular at present are sensors based on the conduction of semiconductors such as tin oxide (Chiba, 1990), or polymers such as polypyrrole (Hierlemann et al., 1995). More sensitive are sensors that 'weigh' impinging molecules: such as piezoelectric crystals (Barko et al., 1995), surface acoustic wave devices (Hivert et al., 1994). More sensitive still is the biological nose. Recently there has been a renewal of interest in optical sensors incorporating fluorescent molecules (Dickinson *et al.*, 1996). The potential uses of these devices go beyond simple automated taste testing, to raw material evaluation, early diagnosis of problems on the production line, health and safety checks and security. In the coming years companies will have a range of products to choose from. These sensors will incorporate the latest in nanotechnology, intelligent response, interaction with on-line management systems, and telemetry.

Traditionally, analytical chemistry has been done in the laboratory, with samples delivered to the door and results duly returned, after a suitable delay, to the waiting engineers and managers. Research in analytical chemistry focused on building bigger and better machines that could analyse ever-smaller amounts of the required analyte. Today there are a number of techniques that can analyse at the ppt (parts per trillion, or 1 microgram in 1 kg) level. Many of the scares that revolve around the dioxin family stem from our ability to analyse it at these low levels. Without analytical chemistry, ignorance could well be bliss.

While this approach to analysis is still important, we have come to realise that many problems are wanting not more accurate analysis to ever lower levels, but fast analysis where it is needed — on line, with results available instantly to those who make decisions. This has required the analyst to leave the security of the laboratory and take to the factory floor.

There are many approaches to designing portable sensors; we might try to scale down our big instruments, or take traditional methods and miniaturise them until they are suitable for our purpose. Colour spot test kits are a good example of this. An alternative way we have taken is to look at that doyenne of portable chemical analysis, the human, and attempt to reproduce some of the functionality of the sense organs in an electronic apparatus. This paper will focus on strategies for monitoring chemicals in the air, and will look forward to applications that will become possible as the technology becomes available.

ELECTRONIC SENSORS COMPARED TO THE HUMAN NOSE

Table 20.1 gives a specification for the human nose, and present (and ideal) electronic noses. A comparison between the human nose and a typical chemosensor reveals we have a long way to go with our technology, but already some of the advantages of non-biological approaches are being realised.

Table 20.1
Comparison between a human nose and an artificial sensor

Property	Biological nose	Electronic chemosensor specification
Number of sensing elements	>1 000 000	<100
Number of possible smells	~10,000	Potentially infinite Practically < 100 per device
Mechanism	Molecular recognition	Chemical reaction/interaction
Transduction	Complex cascade of biochemical reactions	Electrical/spectroscopic with electronic amplification
Processing	Brain	Chemometrics/ artificial neural nets/ pattern recognition
Size	2 x 2 x 5 cm + processor	20 x 20 x 20 cm + computer Often larger but possibility for very small devices via nanotechnology
Performance		
Detection limit Selectivity	1–100 molecules (1 in 10^{20}) Can be confused	ppb–sub ppt (1 in 10^{10}) Should be selective but interferences a problem
Stability	Becomes habituated	Stable throughout life
Life time	~ 70 years	6 months
Precision : reproducibility	Variation between individuals	Good reproducibility under controlled conditions
repeatability	Good only if trained	Good
Linearity	Highly non-linear, Recognition is mainly absent/ present	Linearity through processing calibration possible for quantitative results
Cost	Expensive	Potentially cheap

It is possible to buy a number of commercial instruments for use in industrial contexts. These are often based on the measurement of changes in resistance of a semiconductor, either metal oxide or polymer, which is occasioned by the interaction of a molecule from the gas phase. The chemistry is often low level, as is the measurement of conductivity, and simple pattern recognition is often employed.

Most of these artificial sensors are eclectic in their response to molecules. The nose uses high molecular weight receptor proteins that have been evolved over time to be selective. They are highly sensitive to the shape and electronic configuration of the target molecule. On the other hand, tin oxide, for example, changes its resistivity as a molecule is oxidised by the surface oxygen (Chiba, 1990). It therefore senses any molecule that can be oxidised (ie burnt). Some selectivity is obtained by doping other elements, but unless the environment is highly specified we cannot rely on this sensor to give unambiguous information. A likely development in the next generation of devices is to provide a better level of selectivity, while retaining the advantages of a pattern of responses across an array of sensors (Neaves and Hatfield, 1995).

The advantages of electronic devices lie in their potential to be tailored to respond to any molecule, not just the odorous ones, and to do it repeatably and accurately. In the next section the design and operation of an array of tin oxide based sensors, for monitoring ripening fruit, is described.

A PORTABLE GAS MONITOR FOR RIPENING FRUIT (BOWLES, 1993)

Experimental

The device incorporated two Taguchi sensors (Chiba, 1990) (Figaro Engineering Inc., Osaka, Japan), for example TGS813 and TGS822, in a compartment over which air was pumped at 1 L min^{-1}. The voltage output of each sensor was displayed on a liquid crystal screen, and was also passed to the serial port of a notebook computer (Apple Powerbook) via a 12-bit analogue to digital converter. The system was powered by a 9V rechargeable battery. The device weighed less than 2 kg and was about the size of a notebook computer.

The device was calibrated for ethene and ethyl acetate vapour (10–1000 ppm), and used to monitor the ripening of harvested tomato, rock melon, lemon, banana, apple, kiwi fruit and peach. The fruit was placed in a polythene bag and allowed to equilibrate at room temperature. The air in the bag was then sampled and the output of the sensors monitored with time.

Results and discussion

Figure 20.1 shows a compilation of traces from a lemon over a period of seven days as the fruit progressively ripened. The response of TGS825, calibrated for ethene, shows a peak as the accumulated gas is removed from the bag around the fruit.

With only a single sensor, the increase in evolved gases as a fruit ripens and senesces cannot distinguish between the ethene produced in the early stages of ripening, and the volatile esters found in mature fruit. A second sensor gives the necessary added information. Figure 20.2 is a graph of the peak heights of TGS813 and TGS822 exposed to a ripening banana, taken each day for 14 days. Each sensor responds to ethene and ethyl acetate, but to a different extent. A difference plot thus shows a divergence when the early ethene production gives way to esters.

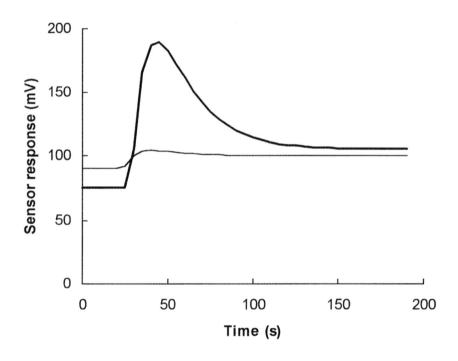

Figure 20.1

A compilation of responses from the Taguchi sensor TGS825 sampling a lemon when just ripe (light trace) and after seven days (bold trace). (Bowles, 1993)

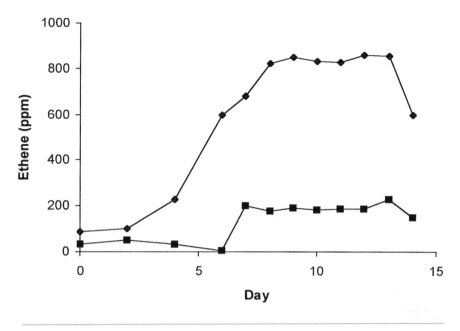

Figure 20.2
The output of TGS813 (upper curve), and the difference between the outputs of TGS813 and TGS822 (lower curve) as a banana ripened. The fruit was sampled once per day and the peak heights converted to an equivalent concentration of ethene. (Bowles, 1993)

Pollution monitoring

The device described above was also developed to measure unburnt petrol and carbon monoxide from automobile exhausts. To illustrate its use, it was driven through the newly opened Sydney Harbour Tunnel. The build-up in signal towards the end of the 1.5 km tunnel coincided with observable smoke and fumes (Figure 20.3). Although the absolute levels of carbon monoxide were always below dangerous limits, the continuous reading that was available from the portable device gave a clear indication of the incorrect setting of the ventilation system.

REQUIREMENTS FOR THE FOOD INDUSTRY

Considering the high cost of present commercial instruments (A$100 000 can be typical), there should be an opportunity for the introduction of cheaper, more targeted systems onto the market. There are already personal monitors used, for example, for testing air in mines for methane, but there is still a need for cheaper, portable and intelligent sensors for a wider range of industrial applications. Table 20.2 gives specifications of a device that would find a ready market, and the present availability in research or commercial instruments. Table 20.3 gives an indication of the types of application that could be viable if such detectors were available. The cheapness of the sensors should allow arrays to be used for a number of applications that are not presently envisaged.

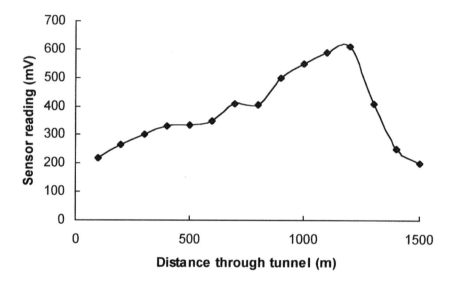

Figure 20.3
The response of a sensor while taken through the Sydney Harbour Tunnel at 5 pm on Wednesday, 2 September 1992. (Bowles, 1993)

Table 20.2
Characteristics of an ideal sensor

Property	Parameters	Available now?
Size	Probe: pen size	Technically feasible
	Computer and telemetry:	Not fully
	small notebook	functional system
Selectivity	Highly selective for	Not for all systems,
	~ 5 specified molecules	but not far away for many
	in given environment	practical problems
Lifetime	Probe at least I year	No
Detection limit	As necessary	Yes, except for trace
		environmental problems
Linearity	Should be able to be	Some better than others
	calibrated across range	
Software	To perform all calibration	No
	and recognition tasks for	
	multiple probes, plus	
	performance monitoring,	
	plus telemetry and intelligent	
	response (alarms, feedback to	
	processes etc.)	
Cost	Probe < $1000	No
	Initial system < $10,000	

Table 20.3
Uses for chemical sensors in industry

Field	Uses
Manufacturing:	
Raw material	Testing against specifications, ripeness of foods
Process	Process control with feedback, identifying bad intermediates at early stage
Product	Testing against specifications
In products	Sensors in refrigerators to detect bad food
Retail	Testing against specifications, adulteration, mislabeling
Agricultural	Monitoring chemical agents (insecticides, herbicides), ripeness, infection (fungal odours)
Health and safety	Environmental monitoring and control of workers Breathalysers for alcohol and drugs Environmental mapping around plant
Security	Sniffing for intruders, entry checks, airport security (bombs, drugs), quarantine

CONCLUSION

The day of the ubiquitous sensor has not come, but industries should already be looking ahead. Research is producing systems that could be commercialised, and future generations of devices will be cheaper, more intelligent, and more useful to industry.

REFERENCES

Barko, G., Papp, B. and Hlavay, J. (1995) Application of pattern recognition and piezoelectric sensor array for the detection of organic compounds. *Talanta,* 42, 475–482.

Bowles, K.C. (1993) Commercial applications of the Taguchi tin oxide gas sensor, BSc Hons. Thesis, University of New South Wales, Sydney.

Chiba, A. (1990) Development of the TGS gas sensor. *Chem. Sens. Technol.,* 2, 1–18.

Dickinson, T.A., White, J., Kauer, J.S. and Walt, D.R. (1996) A chemical detecting system based on a cross-reactive optical sensor array. *Nature,* 382, 697–700.

Gardner, J.W. and Bartlett, P.N. (1994) A brief history of electronic noses. *Sens. Actuators,* B18, 211–220.

Hierlemann, A., Weimar, U., Kraus, G., Schweizer-Berberich, M. and Goepel, W. (1995) Polymer-based sensor arrays and multicomponent analysis for the detection of hazardous organic vapors in the environment. *Sens. Actuators,* B26, 126–134.

Hivert, B., Hoummady, M., Henrioud, J.M., and Hauden, D. (1994) Feasibility of surface acoustic wave (SAW) sensor array processing with formal neural networks. *Sens. Actuators,* B19, 645–648.

Hodgins, D, and Simmonds, D. (1995) The electronic nose and its application to the manufacture of food products. *J. Autom. Chem.,* 17, 179–185.

Neaves, P.I. and Hatfield, J.V. (1995) A new generation of integrated electronic noses. *Sens. Actuators,* B27, 223–231.

PROBABILITIES AND POSSIBILITIES: ON-LINE SENSORS FOR FOOD PROCESSING

D. BARNETT

INTRODUCTION

One of the most significant improvements in the food industry during the next few years is likely to be the development of intelligent processing systems incorporating automated and rapid chemical analysis. By this means product quality and production efficiency will improve and confer an all-important competitive advantage on the products of those manufacturers who use the most appropriate technologies (Barnett, *et al.*, 1992).

Traditionally, the food industry has used very few sensing devices, and these have generally been limited to the measurement of temperature, pressure and perhaps pH. The increasing attention being paid to process efficiency, quality control and effective use of material and equipment means that the older food processing methods, based upon historical experience, are being increasingly supplanted by methods which require automatic or robotic operation (Rymantubb, 1995).

ELECTRONIC NOSES

Electronic noses (Gardner and Bartlett, 1994) offer the food and packaging industries a method of rapid chemical analysis for the improvement of production efficiency and quality control, by classifying complex volatile mixtures during processing and storage (Schaller *et al.*, 1998). Current analytical methods, such as gas chromatography and mass spectrometry, are unsuitable or unable to satisfy the requirement for rapid, simple operation which is demanded in the factory situation. Electronic noses offer competitive advantages to food manufacturers by reducing overall costs and improving product quality in areas such as on-line process control, new material rejection, control of storage conditions, detection of safety problems, taints, and matching products to specification (Hodgins and Simmonds, 1995).

At present, neither commercially produced nor prototype electronic noses are adapted to on-line monitoring or process control. With the possible exception of optical sensors, all the devices suffer from humidity effects, require temperature compensation and need careful control of sampling conditions. Some sensors have saturation problems and show slow recovery to baseline condition. Their sensitivity and speed of operation leaves much to be desired. In addition, demonstrated software and data analysis techniques of the commercial systems are not generally suitable for feedback or alarm situations.

In the current state of the art, the electronic nose to some extent mimics the human sense of smell; but humans have a sensitivity and response repertoire greatly exceeding electronic noses. However, electronic noses can respond to certain chemicals that humans cannot smell, or dare not smell for reasons of safety. Electronic noses are therefore more likely to complement humans and existing technology (Siegmund and Pfannhauser, 1998), than to replace them in many applications.

It is apparent that the existing electronic noses have many promising characteristics and are undergoing further development. However, it is appropriate to consider the particular requirements for an electronic nose or other sensors in a food factory.

The following functions might be served by an adequately performing instrument:

1 Quality check of incoming raw materials Detection of spoilage or deviation from specification could be checked by the supplier as well as the receiving factory.

2 Mixture composition At a mixing point, the mixture of ingredients could be checked by an on-line instrument, to monitor the correct composition of the mixture. A mix

not meeting specification could be diverted before it passed into the next process stage, and corrected. This would be economically advantageous to the factory.

3 Process monitoring At various points in the process of heat transformation or other processing of the material, the product could be monitored by sensors and the information fed back into the process machinery. The result would be savings in operator time and greater consistency of final product (Namdev, et al., 1998).

4 Packaging Packaging of the finished products provide opportunities for monitoring and checking quality against specification. Packaging material sometimes has odour taints from the material itself, or from the inks used. These could be checked by incorporation of an electronic nose or specific sensors into the highly automated packaging machines, thereby providing quality control at the critical point. The result would be minimised complaint from consumers and lower risk of product recall.

5 Shelf-life Shelf-life could be monitored using smart sensors incorporated into sample packages.

Can the present generation of sensors and/or electronic noses address the above situations? If not, is a new instrument required, or is it possible to adapt/modify existing systems to be satisfactory?

Table 21.1 summarises the properties of the chemical gas or vapour sensors being used in sensor arrays, and Figure 21.1 illustrates these sensor technologies.

In principle, biosensors could also be used as sensor elements in arrays but at their present stage of development are not suitable. At present, the vast majority of biosensors operate only in aqueous environments and are not useful as gas sensors.

SINGLE SENSORS

Many research groups, a number of which have commercialised their work, have concentrated on sensor array technology (Dickinson et al., 1998; Doleman et al.,1998), while considerably more laboratories around the world have worked on sensors for the detection of single, specific compounds (Edmonds, 1988).

Single sensors divide into two broad groups: biosensors and chemical sensors.

Biosensors

The particular advantage of biosensors lies in the use of a biologically derived receptor, which has a very high selectivity and sensitivity for the target compound (Turner et al., 1997). However, for use outside the laboratory, biosensors suffer from a number of problems. Their main disadvantage is lack of robustness and short operating lifetime. This is a function of the lability of the biological 'front end' and can be improved upon by chemical or physical manipulation but cannot be eliminated at present.

191

Table 21.1
Outline of sensor technologies in use or under investigation (Kress-Rogers, 1997)

Type of sensor	Principle
Conducting Polymer Sensors (CPS) (see Figure 21.1a)	A polymer (polypyrrole, polyaniline, polythiophene, etc.) is deposited between two electrodes. The polymer is conductive and has a specific resistance. On interaction with analytes the resistance of the polymer changes.
Quartz Crystal Microbalance Sensors (QCMs) (Nanto, et al., 1993) (see Figure 21.1b)	Quartz crystal coated with a selective adsorbent. Adsorption of analytes onto the surface changes the frequency of the crystal's oscillation. Selectivity can be 'tuned' by selection of adsorbent.
Surface Acoustic Wave Devices (SAWs) (see Figure 21.1c)	Similar to the QCMs — use a coated oscillator that changes frequency on interaction. Acoustic waves produced on the surface of the device by application of an alternating electric field to input transducer fingers. The propagated wave is received at theoutput transducer. Selectivity can be achieved by use of various coatings. Reproducibility of manufacture is a problem.
Metal Oxide Sensors (MOS) (see Figure 21.1d)	Ceramic sensing element (often tin or zinc oxide) between two electrodes (at 300°C). Interactions between the analytes and the ceramic sensor result in a change in resistance.
Fibre Optic Sensors (FOS) (Walt et al., 1998) (see Figure 21.1e)	Polymer and a fluorescent dye are deposited on the end of an optical fibre. Light is used to excite the dye and the fluorescence is measured through the fibre. The changing light signal is conveyed through a second optical fibre to the detector unit which also contains a monochromator. Interaction between the polymer/dye complex and the analytes changes the fluorescence.

Advantages	*Disadvantages*
The selectivity of the CPS to analytes can be modified by changing: growth parameters, solvent, monomer, anion used. The sensors can be made reproducibly. Signal generation is simple (resistance).	Humidity sensitivity. Temperature sensitivity. May saturate.
Adsorbents may be tailored to application.	Problems with reproducibility of coating procedure.
Adsorbents may be tailored to application. Theoretically very sensitive.	Problems with reproducibility of coating procedure. Signal generation is more difficult to monitor (frequency change). High frequency operation imposes electronic restraints.
Well-proven technology (safety applications). Simple signal monitoring. Prone to poisoning, slow recovery from saturation and difficult to match precisely. Very rapid response and small size.	Generally poor selectivity — this can be improved by doping with metal ions or running at very high temperatures. Lack of sensitivity.

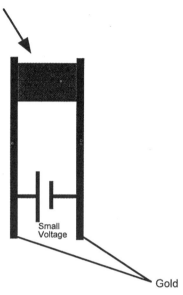

doped conductive polymer is grown as a bridge between electrodes

Small Voltage

Gold

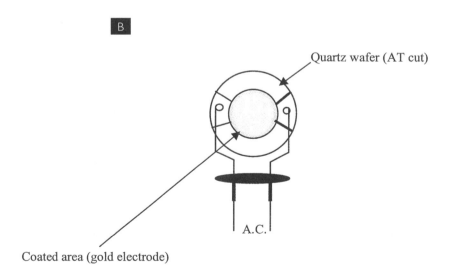

Quartz wafer (AT cut)

A.C.

Coated area (gold electrode)

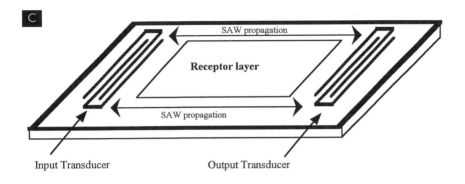

C

SAW propagation

Receptor layer

SAW propagation

Input Transducer   Output Transducer

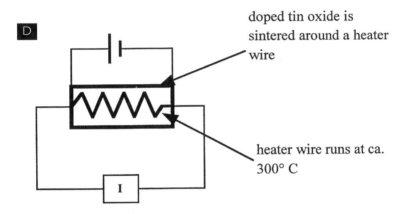

D

doped tin oxide is
sintered around a heater
wire

heater wire runs at ca.
300° C

I

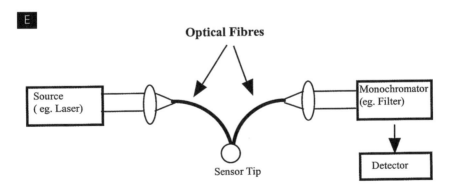

E

Optical Fibres

Source
(eg. Laser)

Monochromator
(eg. Filter)

Detector

Sensor Tip

Notwithstanding the perceived difficulties, biosensors remain a future goal for sensor development in many areas (see Table 21.2). Novel biological molecules may be derived from the olfactory system (Bell, 1996) including odorant binding proteins, olfactory receptors and termination enzymes. These substances offer exciting possibilities for the detection, and possible elimination of odorants, when methods of utilising them are found.

Table 21.2
Examples of some biosensors being developed or used in the food industry

Type of biosensor	Application or detection aim
Enzymic	Amino acids
	Organic acids, sugars, alcohol
	Prototype taste sensor (guanylic acid) for
	Meat freshness
	Fermentation process control
Microbial	Biological Oxygen Demand
Immunosensor	Pathogens, toxic contaminants

Table 21.3
Some food industry applications of chemical and physical sensors (Karube, 1994)

Application	Sensing technique employed
Milk quality measurement and fermentation control (yoghurt)	IR & NIR (fat, protein and total solids) bioluminescence
Moisture in margarine	Microwave attenuation
Quality assurance of baked products	Video image analysis for colour and shape
Nut quality and the rejection of mouldy samples	NIR, X-ray and NMR
Noodle moisture content, shape and thickness	Zirconia moisture sensor, laser camera
Checking of package external appearance and labelling	Very fast optical imaging and 'fuzzy logic'
Vacuum and gas packed meats	Supersonic echoing, leakage detection by He mass spectrometer
Detection of metal object or other foreign body contamination	X-ray, UV, visible light and IR
Oil and fat refining process control	Optical

Chemical and physical sensors

These sensors form the broadest group and have very wide applicability. They cover the detection in foods of such substances as metal ions, anions and organics (including sugars). The techniques shown in Table 21.3 range over electrochemical, optical, conductive, ultrasonic, NMR and RF attenuation. Although many of these techniques could not be considered 'sensory', they are of interest in that they serve to illustrate the range of problems that the food industry must deal with while maintaining quality and efficiency.

CONCLUSION

Great scope exists for the application of knowledge gained from the sensory sciences to many areas of the food industry.

Research areas which are feeding the knowledge-base and may serve in future to meet the challenges facing the application of on-line chemosensors, for the food industry, include:

- manufacture of antibodies to odorants
- development of conductive polymers incorporating bio-polymers
- progress with sensor coatings for specific analytes
- elaboration of data processing using artificial neural nets (ANNs) for sensor arrays
- incorporation of olfactory architecture into ANNs
- development of specific sensor coatings and associated array processing
- investigation of olfactory cell functions for later incorporation of biological molecules from the olfactory and gustatory systems of animals.

Researchers in this area need improved interaction with food industrialists so that both may understand the possibilities suggested by this near-horizon technology, and derive the mutual benefits that are offered.

REFERENCES

Barnett, D., Bell, G.A., Clarke, C.A., Howe, E., Laing, D.G. and Skopec, S. (1992) Food and environmental contaminant detection with novel chemical sensors. In *Proceedings of the First International Conference on Intelligent Materials*. Takagi, T., Takahashi, K., Aizawa, M. and Miyata, S. (eds), Technomic, Lancaster, Penn. USA, pp. 186–193.

Bell, G.A. (1996) Molecular mechanisms of olfactory perception: Their potential for future technologies. *Trends in Food Sci. Technol.*, 7, 425–431.

Dickinson, T.A. White, J. Kauer, J.S. and Walt, D.R. (1998) Current trends in artificial-nose technology. *Trends Biotechnol.*, 16(6), 250–258.

Doleman, B.J. Lonergan, M.C. Severin, E.J. Vaid, T.P. and Lewis, N.S. (1998) Quantitative study of the resolving power of arrays of carbon black-polymer composites in various vapor-sensing tasks. *Anal. Chem.*, 70(19), 4177–4190.

Edmonds, T.E. (ed) (1988) *Chemical Sensors*. Blackie, Chapman and Hall, NY.

Gardner J.W. and Bartlett, P.N. (1994) A brief history of electronic noses. *Sens. Actuators B-Chemical*, 18(1–3), 211–220.

Hodgins, D. and Simmonds, D. (1995) The electronic nose and its application to the manufacture of food products. *J. Autom. Chem.*, 17(5), 179–185.

Karube, I., (ed) (1994) On-line sensors for food processing. *Jpn. Technol. Rev. Section E: Biotechnol.*, 4(2), Gordon and Breach, Switzerland.

Kress-Rogers, E. (1997) Biosensors and electronic noses for practical applications. In *Handbook of Biosensors and Electronic Noses*. Kress-Rogers, E. (ed), CRC, Boca Raton, Florida, pp. 3–39.

Namdev, P.K. Alroy, Y. and Singh, V. (1998) Sniffing out trouble: Use of electronic nose in bioprocesses. *Biotechnol. Prog.*, 14(1), 75–78.

Nanto, H. Kawai, T. Sokooshi, H. and Usuda, T. (1993) Aroma identification using a quartz-resonator sensor in conjunction with pattern recognition. *Sens. Actuators B-Chemical.*, 14(1–3), 718–720.

Rymantubb, N (1995) Computers learn to smell and taste. *Expert Syst.*, 12(2), 157–161.

Schaller, E., Bosset, J.O. and Escher, F. (1998) Electronic noses and their application to food. *Lebensm.-Wiss. Technol.*, 31(4), 305–316.

Siegmund, B. and Pfannhauser, W. (1998) The electronic nose: A new technology for the sensorial analysis of foods. *Ernaehrung (Vienna)*, 22(4), 154–157.

Turner, P.F., Karube, I. and Wilson, G.S. (eds) (1987) *Biosensors — Fundamentals and Applications.*, OUP, NY.

Walt, D.R., Dickinson, T., White, J., Kauer, J., Johnson, S., Engelhardt, Sutter, J. and Jurs P. (1998) Optical sensor arrays for odor recognition. *Biosens. Bioelectron*, 13(6), 697–699.

HOW MACHINES CAN UNDERSTAND SMELLS AND TASTES: CONTROLLING YOUR PRODUCT QUALITY WITH NEURAL NETWORKS

D.C. LEVY AND B. NAIDOO

INTRODUCTION

Machines in the future will use bionic noses (arrays of chemical sensors) to detect and act on complex mixtures of chemical compounds equivalent to what humans know as smells or tastes. Present sensors, however, are not very specific, responding to a range of chemicals, and thus not identifying particular odours very well. The information from the sensors has to be processed to focus on the particular quality of the mixture to be identified. This may be done using conventional statistical techniques. It is attractive, however, to think of teaching the machine to recognise a particular taste or smell, and this can now be done with artificial neural networks (ANNs). By this means, manufacturing processes can be controlled by devices that are trained to respond to a complex chemical environment. The range of applications of ANNs render them a valuable tool for research and industry.

Focusing the sensors using linear regression, or other statistical techniques, is usually successful if the problem has relatively low dimensionality, that is, if the inputs can be combined to give a two or perhaps three dimensional representation of the range of odours.

If the problem is of high dimension (ie the odour is a complex function of many components), these methods may not give reliable results, or the results may not be easy to interpret. Neural networks are then used, as they provide a means to train a network to respond to very complex, non-linear data.

NEURAL NETWORKS

ANNs were designed to emulate neural structures found in the central nervous systems (specifically, the brain) of animals (Rumelhart and McClelland, 1986). Although a variety of ANN architectures exist, the artificial neuron is the basic building block of all types of ANN. The artificial neuron is a simplified model of the biological neuron.

The general structure of the artificial neuron is illustrated in Figure 22.1. The neuron has a set of inputs, each of which is given a *weight*, which can be adjusted. The total *activation* is the sum of all these weighted inputs. This value is then applied to the internal transfer function of the neuron to produce the output, which is sent to other neurons via their weighted connections. Training the neuron is done by adjusting the weights.

The transfer function may be a smooth or a step function, depending on the type of neuron. If *backpropagation* is used to train the network, a smooth function must be used. Backpropagation is a learning scheme where a multi-layer network is organised and trained for pattern recognition using an external teacher. The network outputs are compared to the desired training values and the difference, or error, is propagated back through the layers of the network and used to adjust the network weights according to a training schema.

Multilayer neural networks are composed of several cascaded layers of neurons, which are interconnected, as shown in Figure 22.2. This structure is usually called a Multilayer Perceptron (MLP). It is most common to have three layers. Sometimes more are used, but this can make training more difficult.

Hidden layer sizes can affect network performance quite drastically. If the hidden layer is too small it can impair generalisation, and if it is too large it can overfit the training data, which again causes poor generalisation. (Generalisation refers to the ability of a neural network to use the specific examples with which it was trained, to

Figure 22.1
The artificial
neuron.

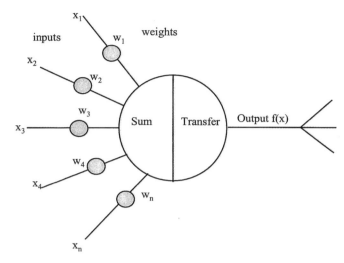

Figure 22.2
Multilayer ANN.

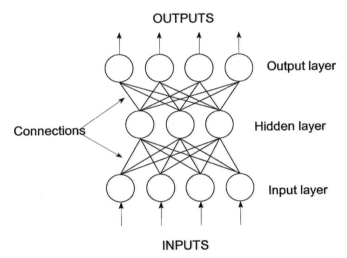

produce sensible outputs to previously unseen inputs.) There are guidelines for determining an optimal hidden layer size, but often it is found by experimentation. In categorisation experiments, where an input must be placed into only one category in the output space, the output layer can be changed to give a binary (on/off, or 'winner take all') indication. This is achieved by seeking the output neuron with the highest output value, and forcing its output to one, while the outputs of the other neurons are restricted to zero.

To train the ANN, a set of training data is needed, which consists of a table of inputs and the corresponding correct outputs. During training, the inputs are applied, the outputs are compared to the correct outputs and the weights adjusted to bring the ANN output 'closer' to the correct output. There are many algorithms for training ANNs (Rumelhart and McClelland, 1986; Warwick, 1995), but only the backpropagation algorithm is described here.

Before training, all the connection weights in the MLP are set to random values. The network is then trained by taking a set of inputs and outputs from the training data, applying the inputs to the network and calculating the outputs. The error for the data pair is the difference between the outputs of the network and the correct output provided by the training data. The error is then propagated back through the network and an error for each layer is calculated. This process is usually repeated for a number of training pairs, and a root mean square error calculated. At the end of a training epoch (sequence of training pairs) the errors for each layer are used to adjust the connection weights so as to reduce the error. The process is repeated cyclically until the network attains some desired performance level.

As the weights are adjusted, the network error (also called energy) changes. This may be expressed graphically, but due to the large number of weights the error surface has a very high dimensionality. The objective of the training algorithm is to locate the global minimum in the energy surface.

Training vectors are presented, in a random order, to the network. This is achieved by randomising the order of the training set, after each sequential pass through it. Since the weight adjustment in each cycle is only marginal, the network gradually moves towards a global solution for the entire training set.

The margin of weight adjustment is determined by a learning rate term. The higher the learning rate, the larger the adjustment margin. If a low learning rate is used, the network takes smaller step sizes on the energy surface, due to small weight adjustments. Therefore, the network is likely to encounter more local minima in the energy surface. Very high learning rates can make the learning process unstable, and often cause the network error to go into oscillation. In such instances, the network takes large steps on the error surface due to large weight adjustments. Usually, high learning rates make the network move rapidly in the general direction of the global minimum. When the general vicinity of the global minimum is reached, the network error oscillates in response to the large step size.

During training, the learning rate may be adjusted for faster convergence on a global solution. Initially, a high learning rate is required, but as the network moves towards the global minimum the learning rate must be reduced to lower values. The lower learning rate reduces the step size and allows the network to locate the global minimum with greater resolution.

The backpropagation algorithm may also contain a momentum term. This technique is used to help the network escape from local minima and speed up convergence. When the weights are adjusted during a learning cycle, the momentum term adds a fraction of the weight change from the previous cycle. The addition of momentum forces the network over small rises and obstacles in the energy surface, towards the global minimum. It also dampens oscillatory behaviour, because the oscillations from cycle to cycle will tend to cancel, and thereby improves the general motion (trajectory) of the network towards a global minimum.

SELF-ORGANISING (KOHONEN) MAPS

Up to now we have been describing *supervised learning,* where the network is provided with the 'correct data'. Self-organising maps (SOMs) are capable of detecting categories within the data, while using an unsupervised training algorithm (Kohonen, 1997). The SOM has an input layer similar to the MLP input layer. Following the input layer, is a two-dimensional Kohonen layer. Figure 22.3 illustrates the SOM.

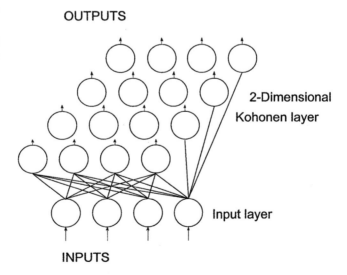

Figure 22.3
Self-organising
map.

OUTPUTS

2-Dimensional
Kohonen layer

Input layer

INPUTS

The neurons in the Kohonen layer are not interconnected with each other. Every neuron in the two dimensional layer receives the same input. The SOM has an unsupervised training algorithm. That is, desired outputs are not specified during training. Therefore, it is not possible to generate an error function.

The SOM clusters similar data into groups called neighbourhoods. These neighbourhoods are physical localities in the two dimensional layer. If an input vector belongs to a particular neighbourhood, then a neuron in that neighbourhood will fire, thus indicating the categorisation. (In general, a neuron 'fires' when its input activation exceeds some threshold defined by the network structure. In a Kohonen network, this is determined by adjacency, measured by the Euclidean distance of each node's weights to the incoming input values.) The location, shape, size and number of neighbourhoods are not predetermined. They develop during the training process, and are determined by the training data, and random state of the connection weights prior to training.

Similar to the backpropagation algorithm, several variations of the SOM training algorithm exist. Before training, all the connection weights are randomised. A cyclical training process then commences. Training vectors are presented to the network at random. After each presentation of a training vector, the input vector is compared with the weight vector of each neuron in the Kohonen layer. The Euclidean distance D_i between the input vector and every weight vector W_i is calculated. The smallest D_i is then found and the corresponding neuron and its immediate neighbours are selected for training. The weight vectors W_i of the winning neuron and its immediate neighbours are marginally adjusted. This is done to reduce the Euclidean distance between these weight vectors and the current input vector. Again, the margin of adjustment is decided by the selected learning rate. Another input vector is selected, and the process is repeated. Eventually, localised areas that correspond to different classes of input develop on the map. These areas are called neighbourhoods and, in this way, inputs are clustered into categories.

APPLICATION OF NEURAL NETWORKS

In a recent research program (Naidoo *et al.*, 1995a and b), several artificial nose prototypes were constructed using a variety of sensor types, and their ability to detect differences in types of food was tested. Experiments have been conducted on cheeses, coffee, and biscuits, to distinguish different types of product and to monitor their quality. One example will be described to show how ANNs can be applied to artificial noses.

An array of Taguchi Gas Sensors (TGS) were used to create a prototype artificial nose to distinguish between several types of cheese. The sensors used have very poor focus, that is, all the sensors in the array responded to all the odours. It was possible to use statistical methods to distinguish between the cheeses over a short period, but the sensors also had other limitations, in that they tended to retain traces of odours to which they were previously exposed, and they drifted.

Initially, an MLP was trained to distinguish between four cheese samples, and it did so with remarkable accuracy. Over time, however, due to high levels of sensor drift, the MLP classifier lost its ability to tell the cheeses apart.

As data were collected over a long period, the effects of sensor drift became very obvious in the collected data. The first step to improve this situation was to vector normalise the input data by mapping it onto a unit hypersphere (a volume representing the distribution of the multi-dimensional input data). Classification rates between 90% and 100% were then achieved. Data normalisation is a common operation and helps to avoid problems with variable data ranges.

At best, linear preprocessing attenuated the effects that sensor drift had on ANN performance, but did not eliminate them. The solution to the problem was sought in the creation of a more robust ANN.

COMBINED SOM AND MLP

A useful approach to the problem of identifying and eliminating unwanted artifacts, like drift, is to cascade an SOM and an MLP, forming a 'cascaded ANN' (CANN). The unsupervised SOM clusters the data, and among the clusters are the artifacts to be removed. The final classification is not performed by the SOM but by the MLP. The architecture is shown in Figure 22.4.

A comparative test was carried out to show the relative performance of the two networks. Data was collected on four consecutive days. The base reference drifted to unique points on each day. Data from the first three days was used for the training set, and data from the fourth day was used for the test set. Test and training data would therefore have different base reference levels. One of the aims of this test was to gauge the network's immunity to base reference drift. All of the data used for the test were vector normalised (mapped onto a unit hypersphere).

During training, each pass through the training set is regarded as a training epoch. At the end of each epoch, the training set classification rate is calculated. To detect possible overtraining (when an improvement in training set classification does not produce improvement in test set classification), the training of each ANN was stopped at three different points and the ANN was tested each time with the test set. A convergence criterion based on training set classification was used to determine when to stop training. The three convergence criteria were 50%, 80% and 95% of training set classification. Training was stopped at the end of the training epoch during which the convergence criterion was achieved by the ANN. Since training was stopped only at the end of an epoch, the actual classification rate usually exceeded the convergence criterion.

The combined networks trained more rapidly than the MLP alone. During the first seven epochs, only the SOM was trained. Thereafter the MLP section was trained and the network needed only three more epochs to achieve the convergence criteria. Furthermore, the standard MLP was prone to overtraining at an early stage. There was no evidence of this phenomenon in the combined ANNs because test and training classification rates improved at each convergence point.

The combined ANNs produced consistently better results than the MLP alone, regularly achieving 95% accuracy and showing robust resistance to the effects of sensor drift.

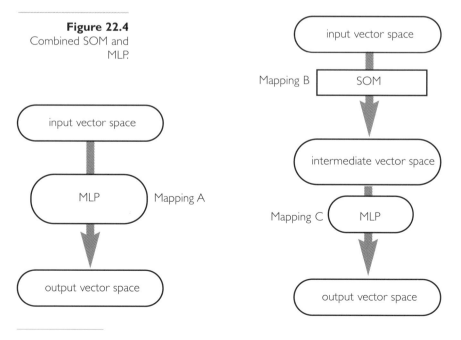

Figure 22.4
Combined SOM and MLP.

CONCLUSION

Cascaded combination of ANNs can drastically reduce training time, and improve performance by reducing the level of non-linearity that each ANN must contend with. The addition of different levels of neural processing eases the burden on the final classification stage (by simplifying the mapping), and allows for improved classification rates and generalisation.

In applying current artificial nose technology to control product quality, it is important to have confidence in the performance of the electronics and to know that imperfections in the sensors can be minimised.

Neural network technology provides a means, not only to train circuits to recognise specific odours, but also to reject sensor imperfections and other artifacts that might otherwise reduce confidence in the results.

REFERENCES

Kohonen, T. (1997) *Self-Organizing Maps.* Springer-Verlag, 2nd ed, Berlin, NY.

Naidoo, B., Levy, D.C., Bell, G.A., and Barnett D. (1995) Classification of data from non-ideal gas sensors. *Sixth Australian Conference on Neural Networks*, Sydney, February.

Naidoo, B., Levy, D.C., Bell, G.A. and Barnett D. (1995) Food odour classification in an artificial olfactory system. *International Conference on Engineering Applications of Neural Networks,* Finland, September.

Rumelhart, D.E. and McClelland, J.L. (1986) *Parallel Distributed Processing: Explorations in the Microstructure of Cognition.* 2 Vols, MIT Press, Cambridge Mass.

Warwick, K. (1995) Neural Networks: An Introduction. In Irwin, G.W., Warwick, K. and Hunt K.J. (eds), *Neural Network Applications in Control, IEE Control Eng Series No 53*, IEE, London, Ch. 1.

INDEX